Recycling Concrete and Other Materials for Sustainable Development

Editors
Tony C. Liu
Christian Meyer

SP-219

Second printing, July 2004

DISCUSSION of individual papers in this symposium may be submitted in accordance with general requirements of the ACI Publication Policy to ACI headquarters at the address given below. Closing date for submission of discussion is October 2004. All discussion approved by the Technical Activities Committee along with closing remarks by the authors will be published in the January/February 2005 issue of either <u>ACI Structural Journal</u> or <u>ACI Materials Journal</u> depending on the subject emphasis of the individual paper.

The Institute is not responsible for the statements or opinions expressed in its publications. Institute publications are not able to, nor intended to, supplant individual training, responsibility, or judgment of the user, or the supplier, of the information presented.

The papers in this volume have been reviewed under Institute publication procedures by individuals expert in the subject areas of the papers.

Copyright © 2004
AMERICAN CONCRETE INSTITUTE
P.O. Box 9094
Farmington Hills, Michigan 48333-9094

All rights reserved, including rights of reproduction and use in any form or by any means, including the making of copies by any photo process, or by any electronic or mechanical device, printed or written or oral, or recording for sound or visual reproduction or for use in any knowledge or retrieval system or device, unless permission in writing is obtained from the copyright proprietors.

Printed in the United States of America

Editorial production: Lindsay K. Kennedy

Library of Congress catalog card number: 2004103240
ISBN: 0-87031-142-5

PREFACE

Recognizing the need to promote and encourage the use of recycled concrete and other materials in concrete construction, ACI Committee 555, Concrete with Recycled Materials, sponsored a two-part technical session on "Recycling Concrete and Other Materials for Sustainable Development" at the 2003 ACI Spring Convention in Vancouver, Canada. Twelve papers were presented by invited experts from Canada, Denmark, Japan, and the United States.

This Special Publication (SP) contains 11 of the 12 papers that were presented at the technical session. The subject areas include the global perspective, challenges and opportunities of concrete recycling, the barriers to recycling concrete in highway construction, and current practices in the European Union, Japan, and USA. This SP also contains research papers on the use of recycled glass as aggregates for architectural concrete, recycled scrap tire rubber, flowable slurry containing wood ash, recycled latex paint as an admixture, crushed stone dust in production of self-consolidating concrete, a new binder using thermally treated spent pot liners from aluminum smelters, and the durability of concrete containing recycled concrete as aggregates that had shown distress due to alkali-silica reaction. Each paper was reviewed by two reviewers in accordance with ACI publication policy.

On behalf of ACI 555, we would like to thank all authors for their contributions and the reviewers for their assistance, comments, and valuable advice. It is our hope that the success of this two-part technical session and the publication of this Special Publication will encourage the use of recycled concrete and other materials in concrete construction, and thereby help our industry to comply with the demands of sustainable development.

Tony C. Liu
U.S. Army Corps of Engineers
Washington, DC

Christian Meyer
Columbia University
New York, NY

TABLE OF CONTENTS

Preface iii

SP-219—1: Recycling Concrete—An Overview of Challenges and Opportunities 1
by E. K. Lauritzen

SP-219—2: Recent Trends in Recycling of Concrete Waste
and Use of Recycled Aggregate Concrete in Japan 11
by Y. Kasai

SP-219—3: Concrete Waste in a Global Perspective 35
by T. C. Hansen and E. K. Lauritzen

SP-219—4: Guidance for Recycled Concrete Aggregate Use
in the Highway Environment 47
by J. S. Melton

SP-219—5: Mitigating Alkali Silica Reaction in Recycled Concrete 61
by H. C. Scott IV and D. L. Gress

SP-219—6: Use of Recycled Glass as Aggregate for Architectural Concrete 77
by C. Meyer and S. Shimanovich

SP-219—7: Properties of Flowable Slurry Containing Wood Ash 85
by T. R. Naik, R. N. Kraus, Y. Chun, and R. Siddique

SP-219—8: Protective System for Buried Infrastructure
Using Recycled Tire Rubber-Filled Cement Mortars 99
by M. Nehdi and A. Khan

SP-219—9: The Use Of Crushed Limestone Dust In Production
of Self-Consolidating Concrete (SCC) 115
by C. Shi, Y. Wu, and C. Riefler

SP-219—10: Recycled Waste Latex Paint as an Admixture in Concrete Sidewalks 131
by M. Nehdi

SP-219—11: Development of a New Binder Using Thermally-Treated
Spent Pot Liners from Aluminium Smelters 145
by A. Tagnit-Hamou and S. Laldji

Recycling Concrete—An Overview of Challenges and Opportunities

by E. K. Lauritzen

Synopsis: This paper consists of an overview of the development of techniques for recycling concrete. Demolition, processing and the recycling of the resulting materials are often analyzed separately. "High quality" recycling of concrete waste does not always correspond to production/use of the product with the highest value, but rather the most feasible product in a specific project or region. It is by analyzing the whole disposal/supply-chain, including the substituted material, that the best effects of recycling can be achieved.

Overviews of methods for environmental evaluations as well as economic considerations are presented. Integrated demolition waste management in Kosovo and an analysis of the potential market in Hong Kong are presented as examples of the worldwide market for recycled materials. Issues regarding the handling of polluted materials will be discussed from a practical point of view. Moreover, some aspects to consider regarding future demolition when producing new concrete products are presented.

Keywords: concrete; construction and demolition waste; demolition; integrated waste management; recycling

2   Lauritzen

Erik K. Lauritzen is managing director of DEMEX A/S, which he founded in 1978. He has worked with the recycling of concrete and masonry since the early 1980s and has been active in the RILEM Technical committees on the demolition and recycling of concrete and masonry. Erik K. Lauritzen has worldwide experience on the demolition and recycling of building materials.

SUSTAINABILITY AND "GREEN" CONCRETE

Since Agenda 21, the Rio Declaration on Environment and Development, was launched in 1992 sustainable development has been one of the key issues of modern society. Some years ago, ACI realized that even if concrete is an environmentally friendly material, Portland cement is the critical component of modern-day concrete. To address this issue and the relationship between sustainable development and concrete technology, the ACI Board of Direction, in 2000, formed a Task Group on Sustainable development and concrete technology. Its mission was to encourage the development and application of environmentally friendly, sustainable concrete materials, design and construction. One of the most important issues of sustainability was the use of recycled aggregates (1).

Fortunately, some ACI members had been far-sighted enough in 1985 to organize Committee 555 – Concrete with Recycled Materials. In 2001 the committee submitted a report *"Removal and Reuse of Hardened Concrete"* (2), which has established a very good basis for the future work of ACI on the sustainability and recycling of concrete.

Parallel to the work of ACI in the USA, the recycling of Construction and Demolition waste (C&D waste) has emerged as a socio-economic priority within the European Union and a considerable amount of research and development has taken place in the frame of RILEM[1]. In 1981 European and Japanese members of RILEM took the initiative to complete the first RILEM technical committee on the demolition and recycling of concrete, including several material research projects in this field. The research of the RILEM technical committees on recycling has been published in proceedings from three international symposia held by RILEM in Antwerp, Belgium 1985 (3), in Tokyo, Japan, 1988 (4) and in Odense, Denmark, 1993 (5). In 2000 the state-of-the-art-report of RILEM Technical Committee 165-SRM on *Sustainable Raw Materials* was edited by C.F. Hendriks and H.S. Pietersen (6).

THE NEED FOR C&D WASTE MINIMIZATION

In all communities it has always been common practice to retrieve valuable materials from the arising waste, e.g. metals and building materials. After some decades in this century with an extensive "use-and-throw-away" philosophy it has been recognized that we cannot continue this uninhibited use of natural resources and pollution of the world

---

[1] **RILEM** : The international union of testing and research laboratories for materials and structures

Recycling Concrete and Other Materials

with waste. It is necessary to change our habits and to revise former common practices within the building & construction industry, as well as within other industries, households etc.

In many countries, industrial as well as developing, C&D waste is considered as harmless, inert waste, which does not give rise to problems. However, C&D waste consists of huge amounts of materials that are often deposited without any consideration, causing many problems and encouraging the illegal dumping of other kinds of waste. Whether C&D waste originates from clearing operations after natural disasters or from human-controlled activities, the utilization of such waste by recycling can provide opportunities for saving energy, time, resources and money (7, 8, 9). Furthermore, recycling and the controlled management of C&D waste will mean that less land is used and better opportunities will be created for the handling of other kinds of waste.

C&D WASTE STREAMS IN THE EU AND USA

A large proportion of C&D waste derives from demolition, rehabilitation, and new construction following normal development, as well as from natural and technological disasters. For example, the production of building materials and goods involves surplus ready-mixed concrete, concrete elements, articles of wood etc., which can be classified as industrial waste.

In the European Union, whose population in 2000 was approximately 370 million, it is estimated that the annual generation of C&D waste is approximately 200- 300 million tonnes - equivalent to ½ - 1 tonne per capita per year in very rough figures (7, 10).

Clear figures regarding recycling do not exist in every EU member country. An EU study calculated that an average of 28 % of all C&D waste was recycled in the late 1990s (11). In the Netherlands 95 % of all C&D waste is recycled and in Denmark 90 % is recycled.

Most EU member countries have established goals for recycling that range from 50 % to 90 % of their C&D waste production, in order to substitute natural resources such as timber, steel and quarry materials. Recycled materials are generally less expensive than natural materials, and recycling in Germany, Holland and Denmark is less costly than disposal (7, 10, 11).

The C&D waste streams in the USA were assessed by Robert H. Brickner in 2002 (12). It is estimated that the amount of C&D waste is 250 – 300 million tonnes per year (2002 figures) and that 20 – 30 % is recovered for recycling (1996 figures).

According to Metha (13) the global concrete industry uses approximately 10 billion tonnes of sand and rock each year and more than 1 billion tonnes of C&D waste are generated every year.

CHALLENGES OF RECYCLING

According to the EU Waste Directive on Prevention and reduction of Waste the key issues are:
- Cleaner technologies
- Recycling, reuse & recovery (R3) of waste
- Management and planning of CDW handling
- Polluter pay principle

This presentation will focus on recycling. C&D waste has a high recycling potential because the majority of it consists of masonry, concrete, and steel. In Europe up to around 90 % of C&D waste consists of concrete and masonry. Buildings erected before the middle of the last century were mainly constructed with masonry, and buildings in the second half of the century were mainly constructed with concrete. The situation in the USA is probably the same.

Based on a global overview (13) it is estimated that the potential of recycling is approximately 50 %, which is equal to approximately 500 million tonnes, which is equal to 5 % of the global consumption of sand and rock.

At present, only very limited amounts of C&D waste are recycled as high-value materials, such as recycled aggregates in new concrete. The majority of such waste is disposed of or used as fill. Since the amounts of C&D waste are constantly increasing, there are many reasons for focusing on methods that promote an increase in the recycling of C&D waste (dumping fees in Europe and the USA are typically from US$ 20-50 per tonne). Present results in Europe show very favorable recycling possibilities in this field.

From a purely economic point of view the recycling of building waste is only attractive when the recycled product is competitive with natural resources in relation to cost and quality. Recycled materials will normally be competitive where there is a shortage of both raw materials and suitable deposit sites.

With the use of recycled materials, economic savings in the transportation of building waste and raw materials can be achieved. In larger recycling projects, such as urban development, renovation of motorways, or clearing of war/disaster-related damages, the total project cost will be dominated by transportation costs. These transportation costs involve the removal of demolition products and the supply of new building materials. In these cases the use of recycled materials is very attractive.

The prospects of systematic recycling of C&D rubble in various parts of the world have been analyzed (7, 14). In order of importance, the three main factors affecting the prospects of recycling C&D waste are
- population size and density,
- occurrence of and access to natural raw materials, and
- level of industrialization.

OPPORTUNITIES FOR RECYCLING

In order to meet the challenges of recycling, it is necessary that all opportunities, barriers and obstacles are detected and considered. The opportunities must be exploited, which can for example involve the recycling of concrete and masonry into aggregate, substituting natural materials:
- Aggregate bound in concrete
- Unbound sub-base and base materials

The opportunities for the production of concrete from recycled concrete are generally described in the ACI 555 report (2), and the recycling of concrete waste in a global perspective is described by Torben C. Hansen and Erik K. Lauritzen (7), Mats Torring (15) and Mats Torring and Erik K. Lauritzen (16).

The overcoming of these barriers and obstacles must be planned and carried out through a long-term action plan combined with adequate research and development. Implementation of the necessary legal, economic and technical instruments requires that initiatives involving legislation and regulations be taken.

Economy

With a condition of market economy, the choice between recycled and natural materials depends on price and quality. The quality of concrete with recycled aggregates can be the same as that of concrete with natural aggregates, but recycled concrete aggregates are regarded with suspicion. Hence, recycled concrete materials will only be preferred where the price of such aggregates is considerably lower than that of the natural materials, even when the recycled aggregates meet given specifications.

Introducing economic instruments, which encourage recycling and the use of recycled materials, can overcome the economic barriers. As an example it should be mentioned that several countries have introduced special taxes and fees in favor of recycling. For example, in 1986 the Danish government introduced a tax on waste that is not recycled but disposed of at landfill sites. Today the tax is DKK 375 (approx. EURO 50) per tonne of waste disposed of at landfill sites.

The major issue of the cost-benefit to society is:
- willingness to pay for the impact on the environment
- willingness to accept the impact on the environment

Policies & Strategies

C&D waste must be considered as a specific individual type of waste associated with the building and construction industry. It is important that the management and handling of waste is carried out by the industry itself. Generally, the building and construction industry is relatively conservative, and changes in normal procedures often take time and need long-term policies and strategies.

One of the most important barriers is the many different interests in building waste.

6   Lauritzen

Usually it is the environmental politicians, departments and public offices who prepare the policies and issues concerning waste recycling and reduction, whereas the building and construction industry is controlled by laws, departments and offices concerned with housing, construction and public works. To co-ordinate the interests of all parties, particularly with respect to the implementation of cleaner technologies in industry, it is necessary that long-term policies and strategies should first have been prepared and implemented.

Danish experience in this field has led to the recommendation that long-term strategies, e.g. for 10 years with respect to achieving goals for the recycling of C&D waste, should be adopted. These must then be continuously revised in accordance with the political situation, and followed up by adequate legislation and regulation at all levels - national, regional and local.

Certification of recycled materials

Demolition and crushing techniques for the production of recycled materials are well known and based on existing technologies. However, some changes in the demolition process, compared with traditional demolition, are required as described below. Even when recycled materials fulfil current standards for natural materials, and even when the prices can compete with the prices of natural materials, certain barriers still exist.

Owing to tradition and psychological barriers the general attitude towards recycling in the building and construction industry is largely prohibitive towards the utilization of recycled materials. Therefore, it is of great importance that recycled materials are officially certified and accepted by all parties in the building and construction industry. For example, in June 1994 RILEM published recommendations for concrete with recycled aggregates (17). A review of international classification and certification of use of recycled C&D waste is presented by Henrichsen (10).

It is recommended that considerable emphasis be placed on specifying the areas of utilization and quality standards for recycled materials. These must be in accordance with the local demand in order to improve confidence in the recycled materials and solve problems regarding the responsibility of using such materials.

Planning demolition projects

A necessary condition for the recycling of building waste is careful sorting of the waste. Waste from new constructions and rehabilitation is sorted either at the production site or at a special treatment site. This separation into materials categories is fairly simple.

The sorting of waste from demolition is, however, a more complicated process. Demolition has until recently been regarded as a low technological process. Rapid demolition and disposal of structures were the main aims of the contractor. Special measures to separate the different types of materials were not possible, owing to the time factor, nor were they desired.

Optimal handling and recycling of C&D waste depends on the materials being sorted in-situ and in co-ordination with the demolition process using demolition technologies and methods as described in the ACI 555 report (2). It is therefore necessary to alter the traditional methods of demolition and introduce selective demolition. This requires that before and during the demolition process an effective sorting of the different materials categories is carried out, thereby preventing any mix of materials leading to pollution of, for instance, recyclable concrete/masonry rubble by wood, paper, cardboard, plastics etc. Since selective demolition takes more time than traditional demolition, detailed planning is considered as mandatory.

It is recommended that demolition projects should be planned and controlled in detail, in the same way as all other building and construction projects, to ensure selective demolition and correct handling of the demolition waste.

Education and information

The most important means to identify and exploit the opportunities and overcome the barriers and obstacles is education and information. It is necessary that the message and understanding of recycling be discussed at technical universities, among private enterprises and public servants.

INTEGRATED RESOURCE MANAGEMENT

In order to achieve the maximum benefit of recycling a management system must be established on a project basis in relation to a specific construction project, e.g. urban development master plan, or on a permanent basis in relation to long-term municipal and C&D waste management system.

The Integrated Resource Management System comprising environmentally and economically balanced management of the following elements:
- Demolition (selective demolition)
- Recycling, reuse recovery
- Handling of hazardous C&D waste materials and non-recyclable materials
- Transportation
- Substituting (saving) natural resources

A presumption of the success of the Integrated Resource Management System is that an effective co-operation between all stakeholders/decision makers has been established in order to avoid a conflict of interests. Conflicts between recycling companies and raw materials companies, for example, could prevent all initiatives towards recycling in general.

The Integrated Resource Management can be implemented according to normal routines of project management in the construction industry, e.g.:
- National policies (legal and fiscal instruments)

8   Lauritzen

- Regional strategies (control of C&D streams, stationary or mobile recycling plants)
- Concepts (high versus low value recycling)
- Feasibility studies (specific proposals for recycling)
- Computer optimization (e.g. waste-resource streams and economic models)
- Master planning
- Design
- Supervision
- Quality & environmental management

GLOBAL VISIONS

The construction industry must aim at durability and sustainability as described by Metha (13). A holistic life cycle based approach is recommended in order to reduce the environmental impact. Further it is mentioned that the resource efficiency of the concrete industry will increase by a factor of five if the service life of most structures built today were 250 years instead of the conventional 50.

Looking at recycling, it is estimated - based on a global C&D waste production of 1 billion tonnes per year - that potentially 50 % should be recycled and could substitute approximately 5 % of the global consumption of sand and rocks.

However, the truth is that there is a long way to go before this level of recycling can be attained in developing countries, and many other environmental problems must be prioritized. But there is no excuse for the industrialized nations not to start the implementation of Integrated Resource Management Systems, aiming of 90 % recycling and maximum substitution of natural resources. There is no doubt that results and experience from European research and development can be transferred to other parts of the world and enable natural (primary) raw materials to be replaced by recycled materials, especially in urban renewal and rehabilitation projects.

CONCLUSION

The development of technologies for the recycling of concrete and the market for various types of recycled concrete materials has proved the viability and sustainability of recycling concrete. The challenges of recycling are dominated by a very high potential of concrete waste all over the world and a demand for recycled materials in order to substitute natural resources. The opportunities for recycling are based on economics, policies & strategies, certification of recycled materials, planning of demolition projects, and education and information.

The success of recycling concrete today in some European countries is based on integrated resource management. The success of recycling in the future is based on global visions for the implementation of recycling concrete worldwide in order to save natural resources and protect the environment.

REFERENCES

1. Malhotra, V.M., 2002, "Introduction: Sustainable Development and Concrete Technology," *Concrete International*, V. 24, No. 7, July 2002.

2. ACI Committee 555, 2001, "Removal and Reuse of Hardened Concrete (ACI 555R-01)," American Concrete Institute, Farmington Hills, MI.

3. Hansen Torben C., 1985, *Proceedings of the EDA - RILEM International Symposium on Re-use of Concrete and Brick Materials*, June 1985.

4. Kasai, Yoshio, 1988, *Proceedings of the Second International RILEM Symposium on Demolition and Reuse of Concrete and Masonry*, Chapman and Hall, Tokyo, 7-11 November 1988.

5. Lauritzen, Erik K., 1993, "Demolition and Reuse of Concrete and Masonry, Guidelines for Demolition and Reuse of Concrete and Masonry," *Proceedings of Third International Symposium held by RILEM*, E. & F.N. Spon, Odense, Denmark, 24-27 October 1993.

6. Hendriks, C.F and Pietersen, H.S. (eds), 2000, "Sustainable Raw Materials – Construction and Demolition Waste (165-SRM)," State-of-the-Art Report of RILEM Technical Committee.

7. Lauritzen, Erik K. and Hansen, Torben C., 1997, "Demolition and Recycling 1986 – 1995," *Danish Environmental Protection Agency, News*.

8. Lauritzen, Erik K., 1994, "Economic and Environmental Benefits of Recycling Waste from the Construction and Demolition of Buildings," *UNEP Industry and Environment*, April-June 1994.

9. De Pauw, C. and Lauritzen, Erik K., 1994, "Disaster Planning, Structural Assessment, Demolition and Recycling," RILEM Report No. 9, E. & F.N. Spon, 1994.

10. Henrichsen, Anders, 2000, "Use of Recycled Aggregates in Europe," presented at Tokyo University, Japan, November 2000.

11. Symonds, Argus COWI and PRC Bouwcentrum, 2000, "Construction and Demolition Waste Management Practice and their Impact," February 2000, DG XI EU Commission.

12. Brickner, Robert H., 2002, "Recycling of Construction and Demolition Waste, Status and Issues in the USA," September-October 2002.

13. Metha, P. Kumar, 2002, "Greening of the Concrete Industry for Sustainable

Development," *Concrete International, July 2002.*

14. Hansen, Torben C. and Lauritzen, Erik K., "Concrete Waste in a Global Perspective," presented at the ACI Spring Convention, Committee 555 April 1st 2003.

15. Torring, Mats, 2001, *Management of Concrete Demolition Waste*, Ph.D. thesis, Norwegian University of Science and Technology.

16. Torring, Mats and Lauritzen, Erik K., 2002, "Challenges of Concrete Constructions," University of Dundee, 5 – 11 September 2002.

17. RILEM Technical Committee TC-121, 1994, "Recommendation for Concrete with Recycled Aggregates," June 1994.

Recent Trends in Recycling of Concrete Waste and Use of Recycled Aggregate Concrete in Japan

by Y. Kasai

Synopsis: This paper addresses first the generation of construction waste in Japan, the extent of recycling, and the shortage of landfill capacity. Second, it chronicles the trend of recycling concrete rubble from 1970 until today. Third, it discusses methods of manufacturing recycled concrete aggregate and introduces the recently developed equipment and methods to produce good aggregate for concrete. Fourth, the problems associated with the use of recycled concrete aggregate are mentioned. Also, the utilization of concrete fines, eco-concrete, and slurries from ready-mix concrete plants are addressed. Land in Japan is very scarce and resources limited. These conditions are similar to those faced by Europeans and Scandinavians. Therefore, this paper should be of interest to those countries and their efforts to build closed-loop material cycles.

Keywords: concrete waste; construction waste; Japan; recycled aggregate; recycled concrete

12 Kasai

Yoshio Kasai is Professor Emeritus of Nihon University. He worked in its Department of Architecture and Architectural Engineering, College of Industrial Technology until his retirement in 1998. He has been closely associated with building materials, especially concrete technology, nondestructive testing, demolition, and recycling of waste materials for construction.

INTRODUCTION

Japan generates about 35 million tons of concrete debris each year. The productive use of this material is very important for the establishment of a sustainable society. This paper addresses the following issues:

1. Generation of construction wastes and their reuse as a resource.
2. Shortage of landfills suitable for industrial waste material.
3. Outline of construction recycling law.
4. Progress made in recycling and reuse of concrete as aggregate.
5. Methods of manufacturing recycled concrete aggregate.
6. Technical Report TR A0006, "Recycled Aggregate Concrete".
7. Problems with quality of recycled concrete.
8. Uses of recycled aggregate concrete.
9. Recycling of concrete rubble for road base and back fill.
10. Use of concrete powder, eco-concrete, and slurries.

The paper mentions the importance of recycling concrete wastes, the methods of producing good recycled aggregate using newly developed equipment and production processes, and problems associated with using recycled concrete in actual construction projects.

GENERATION OF CONSTRUCTION WASTES AND THEIR REUSE AS A RESOURCE

Starting in the fiscal year 1990, the Ministry of Land, Infrastructure and Transport has been conducting nationwide surveys of generation and reuse of construction wastes. The results are shown in Table 1. They show the amount of waste material associated with new construction and demolition, where concrete and asphalt rubble and construction waste wood appear to be due mainly to demolition.

In fiscal year 2000, about 85 million tons of construction waste material was generated, which was 14 million tons less than in fiscal year 1995. The breakdown of this decrease is as follows: 6 million tons of asphalt concrete rubble, 2 million tons of construction mud due to a reduction in public works projects, and the remainder due to reductions in other categories, including the effect of separation and reuse. Fig. 1 illustrates changes in the recycling rate during the ten years, from 1990 to 2000. The recycling rate for concrete rubble has improved from 48% in 1990 to 65% in 1995 and 96% in fiscal year 2000. The recycling rates for asphalt concrete rubble in the same years were 48%, 81%, and 98%,

Recycling Concrete and Other Materials

respectively. For construction mud, the corresponding recycling rates were 8%, 14%, and 41%, mainly due to improvements of plants for dewatering/caking and soil improvement.

The recycling rates for construction wood waste were 31%, 40%, and 38%. This comparably poor showing resulted in spite of recycling of wood as cement-bonded wood particle boards and pulp for paper production. The remainder was used as fuel or incinerated and disposed of illegally.

SHORTAGE OF SAFE LANDFILLS FOR INDUSTRIAL WASTE

Construction waste constitutes about 22% of the 400 million tons of all industrial waste. Landfill sites that remain available for such waste material have decreased year by year, which poses a big social problem. It has become particularly difficult to build new treatment plants in the face of opposition by nearby residents because of the environmental impact. Fig. 2 illustrates the number of permanent landfill sites permitted for industrial waste from 1987 to 2000, while their combined capacities are plotted in Fig. 3 and their remaining life (in years) is indicated in Fig. 4. In fiscal year 2000, the remaining landfill capacity was 176 million m^3, to last for 3.9 years.

Although the remaining landfill capacity decreased and the remaining life increased, the generation of waste decreased over the same time span. Landfill capacity for industrial waste material became so sparse that the government used this opportunity to promote two slogans: "Controlling the generation of waste and decreasing its amount" and "Creating a closed-loop material cycle society".

Fig. 5 illustrates the projected developments of production, waste disposal, and accumulation of building stock during the next hundred years. According to this estimate, the total amount of waste will be four times as great as it is today.

Finally, on May 30, 2002, the Japanese Government decided to enforce the "Law on Recycling of Materials in the Construction Industry, etc", now called the "Construction Recycling Law".

OUTLINE OF THE CONSTRUCTION RECYCLING LAW

The "Construction Recycling Law" is based on the "Law to Promote Effective Recycling of Resources", or "Recycling Law", formulated in 1991 and enacted to further promote recycling of construction related resources. This law refers specifically to concrete rubble, asphalt concrete rubble and construction waste wood.

Conventionally, the owner used to submit to the governor a demolition plan, indicating the type of structure and floor area of the building. However, the new law requires the owner to submit a "Notification Document" for any planned generation of waste in the following cases:
1. Demolition of a building with a floor area greater than 80 m^2.

2. New construction with a floor area greater than 500 m^2.
3. Expansion, repair or remodeling of a building costing more than 100 million yen.
4. Public works projects exceeding a cost of 5 million yen.

It is assumed that each of the last three cases generates the same amount of waste as case 1, which is estimated to be about 30 to 40 tons per house.

The notification document should include various characteristics of the construction, such as the usage, number of stories, projected floor area, and in the case of demolition work, the General Contractor's permit number, registration number of the demolition industry association, etc. In addition, each one of the four cases requires a separate form for filing a demolition plan. Especially in case 1, the method of demolition, separation of waste material, the amount of expected waste generation (quantities of concrete rubble, asphalt concrete rubble, and construction waste wood), the cost of the demolition work, the name of the recycling plant, and the cost of waste recycling should be described. Furthermore, penalties for violations of the law are provided.

PROGRESS IN RECYCING AND REUSE OF CONCRETE AS AGGREGATE

Since the early 1970s, a large number of experimental investigations of the quality of recycled aggregate were undertaken [1]. In 1977, the Japan Building Contractors Society (BCS) published the report, "Proposed Standard of Recycled Aggregate and Recycled Concrete and Commentary" [2]. This was the first systematic proposal for recycled concrete in the world. The standard was presented to the European Demolition Association (EDA) during the RILEM International Symposium on Demolition and Recycling of Concrete and Brick Materials, in Rotterdam, Holland, 1985. Some results of the research for the BCS Standard were also presented on the Second RILEM International Symposium on Demolition and Recycling of Concrete and Masonry. Six papers dealing with properties of recycled aggregate, strength, and durability of recycled concrete were presented [3]. The Building Research Institute of the Ministry of Construction of Japan issued the "Proposal of Quality Standard for Recycled Coarse Aggregate" and the "Proposed Standard for Recycled Coarse Aggregate Concrete" in November 1986. In the latter standard, the issue of recycled fine aggregate followed the BCS Standard of 1977.

Table 2 lists the quality standards in terms of water absorption and soundness of the aggregate. In general, it states that recycled aggregate with larger mortar content leads to high water absorption and low quality.

Although some proposals were made subsequently, the Ministry of Construction published the "Tentative Quality Standard Proposal for Concrete" in April 1994. Table 3 lists design strengths for recycled concrete, where the recycled aggregate categories correspond to those of Table 2. Combining high-class coarse and fine aggregate produces Class I recycled concrete, while a combination of low-class aggregate produces low-class

Recycling Concrete and Other Materials

concrete. Table 4 lists examples of types of mainly civil construction and products produced with recycled concrete according to the abovementioned proposals.

METHODS OF MANUFACTURING RECYCLED AGGREGATE

The manufacture of recycled aggregate has many different aspects. These shall be discussed in detail in this section.

Requirements for Original Concrete for Recycled Aggregate

The original concrete to be used for recycled aggregate has to fulfill the following requirements: 1) It shall be free of harmful components such as soil, mud, asphalt, ALC rubble, etc, and 2) it shall be free of harmful substances such as chlorides and alkali-reactive material. In order to guarantee this, an engineer of the concrete debris-processing plant should inspect the demolition site prior to demolition and search for signs of corrosion of reinforcing steel due to chlorides and concrete cracks caused by alkali-aggregate reaction, before accepting the concrete debris as source material for recycled aggregate. It is proposed that a contract between the concrete debris supplier and the processing plant operator be signed after such an inspection.

Recycled Aggregate Manufacturing Process

The general manufacturing process for recycled aggregate is outlined in Fig. 6. After initial crushing, metal chips, impurities and pieces of wood are removed manually on a sieving screen. The material is then crushed again, and metals are removed in two or three stages by a magnetic separator.

Grain Improvement of Recycled Aggregate

Mortar may adhere to the surface of the original aggregate. Its removal will reduce the amount of recycled aggregate with acute angles. Fig. 7 shows the abrading machine, which was developed by the Nuclear Power Engineering Corporation. Crushed coarse aggregate is thrown from the right-hand gate into a circular tube, where it is forced to the left side. After grinding it with a middle cone and the end rotor cone, the mortar is removed from the surface of the original aggregate. Four repetitions of this process produce a good quality aggregate [4].

Fig. 8 shows a machine in which a number of chains are attached to a turning shaft. When the rough crushed concrete rubble is loaded from the top, it is hit repeatedly by the chains and turned into good aggregate [5]. The machine shown in Fig. 9 has an eccentric rotor in a drum and is therefore called eccentric rotor mill. When the roughly crushed aggregate is fed from the top, the inner drum turns and thereby removes the mortar from the recycled coarse aggregate [6]. A plant for processing high-quality recycled aggregate by the eccentric processing machine is shown in Fig. 10. Fig. 11 compares material before and after processing, with almost all adhered mortar removed.

16   Kasai

Removal of Impurities and Low-Density Aggregate by Rising Water Stream

Fig. 12 shows a plant to separate coarse aggregate by density, using a rising stream and vibration. The coarse aggregate is fed into the left side, screened, and while moving to the right side, washed by the rising water stream. Low-density particles float to the top and are discharged into an exhaust ditch. The remaining high-density particles are raked out from the right side [7]. A similar apparatus to separate fine aggregate by density has also been installed, using the same principle, and produces good-quality aggregate.

Removal of Fines by Air Separator

The air separator separates fine powder from fine aggregate. The fine aggregate is fed into the upper part of the equipment, and a centrifugal force and rising air currents remove the fine powder.

Heating and Grinding Method to Manufacture Recycled Aggregate

After preliminary crushing, the original concrete is heated in a shaft furnace up to 300°C for 40 to 60 min, which produces fine cracks between the cement paste and the aggregate (Fig. 13). It is then crushed and the mortar abraded in a tube mill [8]. Good recycled coarse and fine aggregate can be obtained, with small amounts of mortar adhering to the particles and moisture absorption almost as low as that of natural aggregate. Mitsubishi Materials Corp., with financial support from the Ministry of International Trade and Industry (MITI), has built a pilot plant of 300 kg/hr capacity. Fig. 14 compares original concrete rubble with material that was either heated to 300°C or not heated. The composition of the heated recycled concrete is almost the same as the original aggregate, whereas the recycled aggregate that was not heated produced 44% fine powder. Fig. 15 shows a plant of 3 ton/hr capacity. This plant produces about 30 to 40% of the original concrete, with an increased amount of fine powder [8]. The process requires 29 kWh of electric power and 8 kg of kerosene to produce one ton of concrete [9]. At present, the Nuclear Power Engineering Corporation is undertaking a demonstration project to process a huge amount of concrete debris from a demolished nuclear power plant.

TR A0006:2000 – CONCRETE USING RECYCLED AGGREGATE

The technical report "TR A0006 – Concrete using Recycled Aggregate" was issued in November 2000, in the name of the minister of the MITI, assuming that it will become a JIS Specification in the future. The report specifies the minimum quality requirements for recycled aggregate, Table 5. This table also lists the permissible amounts of fines. Since the quality of recycled aggregate fluctuates considerably, quick test methods are required. It is believed that the rate of water absorption of coarse aggregate is an indicator of quality, therefore it is stated that recycled aggregate with a large amount of mortar exhibits a high rate of water absorption, which points to low quality. The test method, described in the appendix of the report, is as follows.

Method to estimate water absorption and loss of soundness of recycled coarse aggregate.

This test method makes use of the close relation between the rate of water absorption and the amount of mortar attached to the original aggregate. After determining the 100 kN crushing value by the BS 812 method, the water absorption Q, in percent, can be estimated by the formula,

$$Q = 0.85 \, C_g + 1.50$$

where C_g is the 100 kN crushing value, in percent [10]. The loss of soundness, P, of recycled coarse aggregate, also in percent, is then estimated according to,

$$P = 8.0 \times Q$$

Thus, for a Q-value of 5%, the soundness value P becomes 40%. According to Table 2, the durability of second-class and third-class coarse aggregate may be insufficient, but definite conclusions should be drawn only after further evaluation.

Method to estimate water absorption of recycled fine aggregate.

Because of the relationship between water absorption and unit mass of recycled fine aggregate (Fig. 16), the following formula can be used to estimate the water absorption of recycled fine aggregate, Q, [11],

$$Q = -25 \, T + 41$$

where T is the unit mass of the recycled fine aggregate (kg/l – should be converted to %). This formula requires further examination.

Fines in recycled aggregate.

If the recycled aggregate contains an excessive amount of fines, the amount of required water and shrinkage increase, and the freeze-thaw cycle resistance decreases. The upper limit for fines in recycled coarse aggregate was specified as 2% and for recycled fine aggregate as 10% (see Table 5). If washed with clean water in a machine as shown in Fig. 12, the coarse aggregate can easily meet this requirement, and with the help of an air separator, also fine aggregate can meet this limit.

Salinity of recycled aggregate.

The TR A0006 report cautions against an excessive chloride content in the mortar attached to recycled concrete aggregate. After discussion with the producer, the buyer should specify an upper limit for such chloride content. Unless the recycled aggregate is to be used in unreinforced concrete structures, a visual inspection prior to demolition of the building should be conducted as stated previously, and the specified limit of 0.3 kg/m^3 of chloride should be kept.

Alkali-aggregate reaction of recycled aggregate.

If there is the risk of alkali-aggregate reaction with the recycled aggregate, portland blast furnace slag cement Class B can be used. Report TR A0006 specifies that such cement or portland fly ash cement Class B can be used.

PROBLEMS WITH QUALITY OF RECYCLED CONCRETE

The problems with the quality of recycled concrete have already been addressed by numerous papers. The strength of recycled concrete is usually lower than that of normal concrete, for the same water-cement ratio. This fact is particularly noticeable if the original concrete was weak to begin with, because it contained a considerable amount of weak mortar.

Larger shrinkage.

Since recycled concrete contains the mortar of the original concrete and the fines resulting from the crushing operation, the shrinkage of recycled concrete tends to be high.

Low freeze-thaw cycle resistance.

Many reports show that recycled concrete has poor freeze-thaw cycle resistance. Whereas some experiments with air-entrained recycled concrete gave good results, S. Nagataki [12] showed that the freeze-thaw cycle resistance of concrete made with recycled aggregate from non-air-entrained concrete was low. If the original concrete was air-entrained, the results were sufficiently good. It is necessary to confirm these results, because the influence of the amount of fines is significant. According to Report TR A0006, "it is difficult for recycled concrete to have good freeze-thaw cycle resistance. Therefore recycled concrete should be used in underground applications and in regions with mild climate."

Mixed use of recycled and regular aggregate.

When substituting regular aggregate for recycled aggregate, a 30% replacement value results in almost the same strength as achieved with regular aggregate alone, and the effects on the other properties are also small. With 100% recycled coarse and fine aggregate, the strength and durability of the concrete are decreased significantly. Depending on a particular application, there are suitable combinations of recycled and regular aggregate. For example, there is the possibility of combining 100% recycled coarse aggregate with 100% regular sand, or 50% each of recycled and regular coarse and fine aggregate, etc.

USE OF RECYCLED AGGREGATE CONCRETE

Recycled concrete has been studied for more than 30 years, and there exists an extensive literature. However, it was not easy to put the findings to practical use [13]. The

Recycling Concrete and Other Materials

following examples of applications in real structures are particularly noteworthy.

In 1984, the Building Research Institute of the Ministry of Construction built two small test structures using recycled coarse aggregate concrete. Because the recycled aggregate was processed very carefully, no deterioration was observed at all.

In 1995, about 3000 m^3 of recycled concrete was placed as the foundation for a temporary structure for the International City Exposition in Tokyo, but the plan was cancelled and the concrete removed. This was very unfortunate for the recycled concrete community.

In 1996, lodgings for members of the House of Counselors were constructed as a demonstration project. 165 m^3 of recycled concrete were placed for a 6 cm thick flat slab with an area of 2750 m^2.

In 1997, two cast-in-place concrete piles with a diameter of 1 m and a length of 30 m were constructed with 100% recycled concrete. Cores with 10 cm diameter were taken from the central part to the full depth of the two piles, and density, compressive strength, etc. were determined. The test results were not bad. At this site, more than 20 piles of the same size were constructed for a five-story building. The building is performing without problems. Since cast-in-place concrete piles are subject to neither drying nor freeze-thaw cycles after construction, they are an application where recycled concrete can be used with advantage [14].

In 1998, the Housing and Urban Development Corporation built an experimental two-story building of 144 m^2 and a total floor area of 203 m^2. To date, the building portion built entirely with recycled concrete has shown no abnormal behavior, although net-like cracks were observed. The cracks are still growing gradually, but remain within permissible range. Moreover, the HUDC utilizes concrete debris obtained during the demolition of a housing complex for a rainwater filtration system, a container for crushed aggregate storage, a small retaining wall, as prepacked material for a pavement, and as a soil and ground improvement material.

In 2000, a single-story pump room with an area of 127 m^2 for a water tank was built using recycled aggregate that was produced by the heating and abrasion method. The recycled concrete can be used in the same way as normal aggregate concrete.

Recycled concrete is also being used as a porous material for plant beds. Some companies have launched trial production of road curbstones and other simple concrete products like blocks and others for which there is a public demand. However, the public demand was very low.

In Japan, there are only three or four ready-mix recycled concrete plants. Their output is used mainly as foundation for wooden houses and exterior concrete, etc. Fig. 17 shows the volume of ready-mix recycled concrete delivered by the Tateishi plant. About 40,000 m^3 of recycled concrete is produced each year. Therefore, about 120,000 m^3 of such concrete is produced altogether in Japan each year.

In recent years, test results for bending, shear and bond strength of small recycled concrete beams were reported. According to these reports, their mechanical properties are not much different from those of normal aggregate concrete beams.

RECYCLING OF CONCRETE RUBBLE FOR ROADBASE AND BACKFILL

As shown in Table 1, 35 million tons of concrete rubble were generated in fiscal year 2000. The recycling rate was 96%. At the demolition site, the concrete rubble is crushed by mobile crushing equipment and then used on site as road base material, for land reclamation, backfill, etc. The remaining concrete rubble is sent to a processing facility, where the majority of the material is crushed to minus 40 mm size. Table 6 lists the main uses of concrete rubble according to the "Recycling Law". The main uses today are as crushed stone layer and recycled fine aggregate. In 1992, the Japan Road Association issued a "Technical Guideline for Recycled Pavement Construction" for the use of concrete rubble as road base material.

USE OF CONCRETE POWDER, ECO-CONCRETE AND SLURRY

Use of Concrete Powder.

When crushing concrete rubble to produce recycled aggregate, a large volume of fines is generated, which contains considerable amounts of cement hydration products. Many experiments have been carried out to determine whether this powder can be used as cement replacement, but it was found not to contribute to strength development.

In 1976, the author has tested a powder made entirely from concrete rubble. The molar ratio was adjusted by adding lime. After mixing the powder with water, it was subjected to autoclave curing, immediately after which strengths of 30 – 40 N/mm^2 were obtained. The strengths changed as the concrete mixes were varied and different original concrete rubble was used. The reason was probably due to the mineral composition of the aggregate for the original concrete, which depends on the place of production [15].

In another test, the fine powder collected by the air separator was heated up to 600°C and ground to an even finer powder in a mill. It was possible to replace about 10% of the cement [16]. The fine powder can be added as a filler to highly flowable concrete or to control segregation. However, as more water is added, the strength decreases and shrinkage increases.

Mixed with mud and chemical admixtures, the fine powder is being studied as a material to fluidize soils. If the powder is produced near a cement plant, it can be readily used as a raw material for cement production. But this has not yet been put to practice.

Recycling of limestone aggregate concrete (Eco-Concrete).

According to Tomosawa et al. [17], concrete rubble can serve as raw material for cement production, if limestone aggregate was used in the original concrete. This material has been named "Eco-Concrete" and is currently undergoing detailed study. However, if the use of limestone aggregate is as widespread as in Japan, it will be depleted within 40-50 years, and no cement plants will be operational when Eco-Concrete structures are demolished, because the service life of reinforced concrete structures is 60 years or longer. Therefore, it may be said that Eco-Concrete is not reusable after all. The use of limestone aggregate cannot be recommended. On the contrary, it should be reserved for more important applications such as for cement manufacture, steel production, and as raw material for many chemicals.

Reuse of concrete slurry.

Research on the reuse of washing water from ready-mix plants, from returned concrete, and sludge removed from concrete has started in Japan in the 1970s. The JCI has compiled a report in 1975. The treated water can be used as mixing water for concrete. Furthermore, the JCI committee determined that up to 3% sludge (converted to dry equivalent) can be used for new concrete (not as cement replacement). This recommendation was accepted in JIS A-5308 [18].

Since civil engineers have been concerned in recent years about the durability of concrete, the idea of mixing sludge with new concrete is not very popular. But the disposal of sludge has become a major problem. It can be processed into a cake, and if a cement plant is nearby, used as a raw material for cement production. After drying and heating, and mixing it with ground granulated blast furnace slag, it can be turned into a blended cement. Although such a material has been produced temporarily, it is not being manufactured at present.

CLOSING WORDS

This paper dealt with recent trends in industrial waste processing, recycling of concrete rubble and related waste materials in Japan. The main points made were as follows.

1. The amounts of construction waste and recycling as a resource were mentioned. More than 95% of concrete rubble are being recycled and used mainly as road base material.
2. Landfill sites suitable for industrial waste are becoming scarce. Their combined capacities will last for only another 2 to 3 years.
3. The Ministry of Land Infrastructure and Transportation has enacted a "Construction Recycling Law" in May, 2002, to enforce the recycling of construction waste.
4. Methods of producing recycled aggregate were discussed, and new technologies to produce high-quality aggregate were introduced.

5. An outline of Technical Report TR A-0006 (Recycled Aggregate Concrete) was given.
6. The trends of recycled concrete usage from 1984 to date were outlined, which at present results in about 120,000 m^3 of concrete being produced annually.
7. Problems associated with "Eco-Concrete", fine concrete powder, and concrete slurries were mentioned.

REFERENCES

[1] Kasai,Y and Yamada,T, "Experiments on Recycled Aggregate Concrete", Proc., Annual Meeting of the College of Industrial Technology, Nihon University, 1973 (in Japanese).
[2] Committee on Reuse of Construction Byproducts, "Standards for Use of Recycled Aggregate and Recycled Concrete (Proposal) and Commentary," Building Contractors Society (BCS), May, 1977 (in Japanese). This standard was presented to EDA–RILEM International Symposium on "Demolition and Recycled Concrete and Brick Materials", Rotterdam, 1985 (in English).
[3] Kasai,Y, Ed., Proceedings of the Second International RILEM Symposium on "Demolition and Reuse of Concrete and Masonry", Tokyo, Vol. 2, Reuse of Demolition Waste, Chapman and Hall, 1988.
[4] Sakazume,Y, Matsuo, K et al., "Development of Production Technique for High Quality Recyclable Aggregate, Part 4, Method of Screw Grinding". Summaries of technical papers of annual meeting, A.I.J, 2002, pp.1027–1028 (in Japanese).
[5] Takesita, H, Matsubara, S. et al, "Study on Recycled Aggregate from Waste Concrete", Proceedings of the JCI, Vol. 24, No.1, 2002, pp. 1323–1328.
[6] Yonezawa, T, Kamiyama, Y et al, "Study on a Technology to Produce High Quality Recycled Coarse Aggregate", J. Soc. Mat. Sci., Vol. 50, No.8, 2001, pp.835–842 (in Japanese).
[7] Hayashi, M, "Refining System of Recycled Aggregate by Lasa Water Separator", Proc., Regular Meeting of Japan Institute of Aggregate Technology, pp.17-20, Oct. 1995 (in Japanese).
[8] Shima, H, Konosu, K et al, "Development of Recovering Technology of High Quality Aggregate from Demolished Concrete with Heat and Rubbing Method", Proceedings of JCI, Vol.22, No.2, 2000, pp.1093–1098 (in Japanese).
[9] Shima, H, Tateyasiki, H et al, "Life Cycle Assessment of High Quality Aggregate Recovered from Demolished Concrete with Heat and Rubbing Method", Proceedings of JCI, Vol.23, No.2, 2001, pp.67-72 (in Japanese). And Recovery Plant Catalog of Mobile Type High Quality Aggregate, Mitsubishi Material Ltd. (in Japanese).
[10] Kato,T, Kawano,H et al, "An Investigation of Simple Quality Estimation for Recycled Aggregate", Proceedings, 50th Annual Conference of Japan Society of Civil Engineers, No 5, 1995, pp. 200–201 (in Japanese).
[11] Yoshikane, T, Nakajima,Y et al, "Recycling Concrete with Zero-Emission Goal", Journal of Concrete Technology, "Resources and Raw Materials", 2000, General Presentation(C), pp.129-132 (in Japanese).
[12] Nagataki, S, "State-of-the-Art Report: Development of New Recycling Method for Construction Material Considering Life Cycle of Construction", 1998 (in Japanese).

[13] Kasai, Y, "Barriers to the Reuse of Construction Byproducts and the Use of Recycled Aggregate in Concrete in Japan", R.K. Dhir, et al, Eds., "Sustainable Construction. Use of Recycled Concrete Aggregate", November 1998, pp.434–444, Thomas Telford.

[14] Yanagi, K, Kasai, Y et al, "Experimental Study on the Applicability of Recycled Aggregate Concrete to Cast-in-Place Concrete Piles", R.K. Dhir, et al, Eds., "Sustainable Construction. Use of Recycled Concrete Aggregate", November 1998, pp.359-370. Thomas Telford.

[15] Kasai, Y et al, "Autoclave Curing of Concrete Powder", Summaries of Technical Papers of Annual Meeting, AIJ (Structural Series) 1976, pp.133-134.

[16] Kasai, Y and Yuasa, N, "Effective Use of Concrete Powder Byproducts from Recycled Aggregate", International Seminar on Recycled Aggregate, sponsored by Niigata University and Japan Concrete Institute, Sep. 29, 2000, Niigata-City, Japan.

[17] Murata, Y, Tomosawa, F and Isohata, T, "Study on Practical Use of Completely Recyclable Concrete", JCA Proceedings of Cement and Concrete, No.51, pp.500–505, 1997.

[18] Kasai, Y, "Recycling Waste Water and Cement Slurry Disposal at Ready-Mix Concrete Plants", Japan-US Scientific Seminar, San Francisco, 1979, pp.101-110.

Table 1. Amounts of Construction Byproducts

Fiscal Year	1990		1995		2000	
Items	Generated ($\times 10^6$ tons)	Recycled (%)	Generated ($\times 10^6$ tons)	Recycled (%)	Generated ($\times 10^6$ tons)	Recycled (%)
Concrete Rubble	25.4	48	36.0	65	35.0	96
Asphalt Concrete	17.6	48	36.0	81	30.0	98
Construction Waste Wood	7.5	31	6.0	40	5.0	38
Construction Waste Mud	14.0	8	10.0	14	8.0	41
Mixture of Construction Waste	1.5	14	10.0	11	5.0	9
Other Waste	1.5	40	1.0	–	2.0	–
Total	75.9	–	99.0	58	85.0	85

Table 2. Quality of Recycled Concrete Aggregate

	Recycled Coarse Aggregate			Recycled Fine Aggregate	
	1st Class	2nd Class	3rd Class	1st Class	2nd Class
Water Absorption (%)	<3	<3	<5	<7	<5	<10
Loss of Soundness	<12	<40*	<12	–	<10	–

* In this case, do not consider freeze-thaw deterioration

Table 3. Design Strength of Recycled Concrete

Class of Recycled Concrete	Coarse Aggregate	Fine Aggregate	Design Strength (MPa)
I	Recycled coarse aggregate 1st class	Ordinary aggregate	>20 (RC)
II	Recycled coarse aggregate 2nd class	Ordinary aggregate or Recycled aggregate	>16 (Plain)
III	Recycled coarse aggregate 3rd class	Recycled aggregate	<16

Recycling Concrete and Other Materials

Table 4. Examples of Structures Using Recycled Concrete (Civil Construction, not Buildings)

Class of Recycled Concrete	Usage	Coarse Aggregate	Fine Aggregate	Structures
I	RC & Plain C	Recycled coarse aggregate I class	Ordinary aggregate	Bridge substructure, retaining wall, tunnel liner
II	Plain C	Recycled coarse aggregate II class	Ordinarye or recycled fine aggregate 1st class	Concrete block, foundation of road attachment, side drain, foundation of water catchment box, fill concrete, side slope frame, gravity bridge basement, wave breaker, erosion control dam
III	Concrete sub slab	Recycled coarse aggregate III class	Recycled fine aggregate 2nd class	Concrete sub slab, concrete screed, back fill concrete, non-structural members

Table 5. Quality of Recycled Concrete Aggregate (After TR A 0006)

	Recycled coarse aggregate	Recycled fine aggregate
Water Absorption (%)	<7	<10
Amount of Fineness (%)	<2*	<10*

* When the manufacturing method is wet process, this value will be satisfied, so the test is not needed. For dry process, this value is sometimes not satisfied, so it should be confirmed by wet test.

Table 6. Main Usage of Concrete Rubble in "Recycling Law"

Category	Main Usage
Crusher Run	Load subbase material, back fill and base material for public construction, basement material for building
Recycled Fine Aggregate	Filling and basement material for construction
Recycled Size Controlled Crushed Aggregate	Other upper layer roadbed material
Recycled Cement Stabilized Roadbed Material	Roadbed material for road pavement and other pavement
Recycled Lime Stabilized Roadbed Material	Roadbed material for road pavement and other pavement

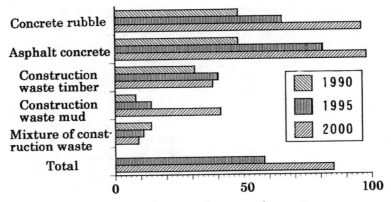

Fig 1: Recycling rate of construction wastes

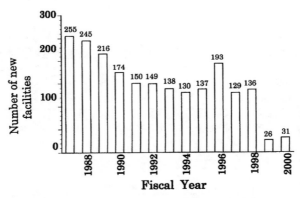

Fig 2: Number of new facilities for industrial wastes

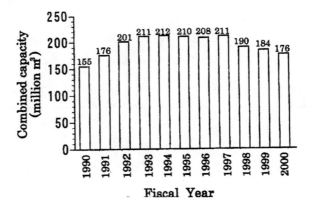

Fig 3: Remaining capacity of safe landfills for industrial waste

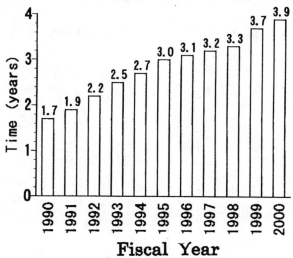

Fig 4: Remaining life of safe landfills for industrial waste

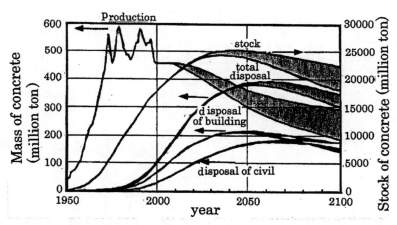

Fig 5: Mass of production, disposal and stock of concrete in the future

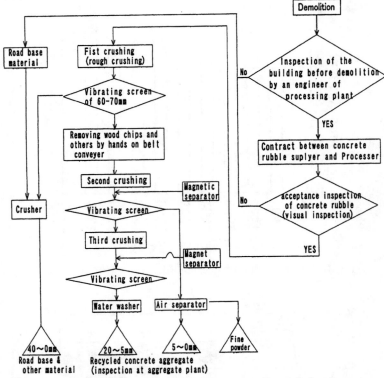

Fig 6: Outline of manufacturing process of recycled aggregate

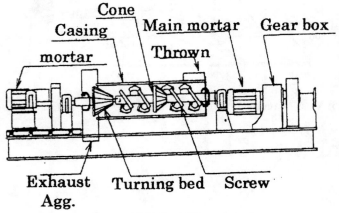

Fig 7: NPEC rubbing method

Recycling Concrete and Other Materials

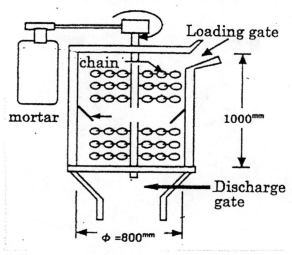

Fig 8: Schematic figure of chain Rotor machine

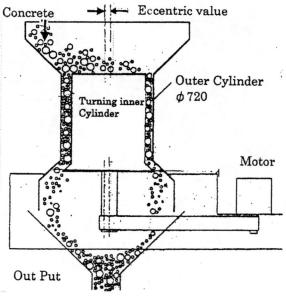

Fig 9: Outline of eccentric rotor processing machine

Fig 10: High quality recycled aggregate processing

Fig 11: Comparison of processing
Before and after

Recycling Concrete and Other Materials

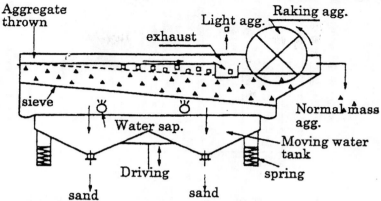

Fig 12: Removal of impurities and low density aggregate by rising water stream machine

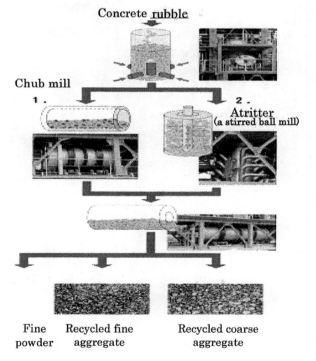

Fig 13: Process of heating and abrading method

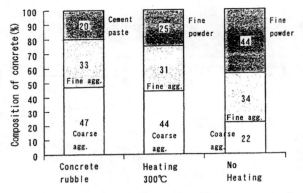

Fig 14: Composition of orginal concrete and recycled concrete

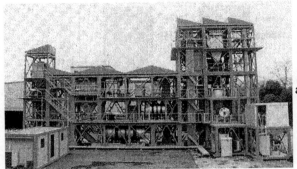

Fig 15: View of heating and abrading plant

Outline of heating and abrading plant 3 t/h capacity

Recycling Concrete and Other Materials 33

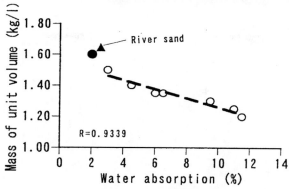

Fig 16: Relation between water absorption of recycled fine aggregate and mass unit volume

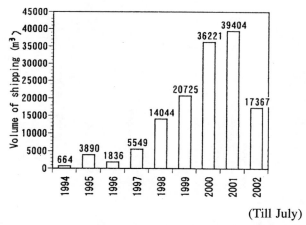

Fig 17: Volume of shipping of recycled ready mixed Concrete in a year

SP-219—3

Concrete Waste in a Global Perspective

by T. C. Hansen and E. K. Lauritzen

Synopsis: The reduction and utilization of waste and by-products is one of many challenges facing mankind in the modern world. This paper consists of an overview of the handling of the world's concrete waste and is concerned with unbound applications of blended construction and demolition (C&D) rubble for pavement bases and sub-bases in road construction. The problems of recycling mixed wastes are considered and global perspectives are presented on the use of C&D rubble. The paper proposes that studies should be carried out in order to research the technical properties of blended C&D rubble, which the authors believe would be economically and environmentally justifiable in promoting the widest possible range of recycled products for road construction.

Keywords: asphalt; bases; bricks; concrete; construction and demolition waste; demolition; recycling; road construction; sub-bases

Torben C. Hansen is Professor Emeritus at the Technical University of Denmark and FACI. He has worked with cement and concrete over the last 45 years and with the recycling of concrete for 20 years. He was Chairman of the RILEM committee on the recycling of concrete from 1980 to 1988 and President of RILEM from 1988 to 1991.

Erik K. Lauritzen is managing director of DEMEX A/S, which he founded in 1978. He has worked with the recycling of concrete and masonry since the early 1980s and has been active in the RILEM technical committees on the demolition and recycling of concrete and masonry. Erik K. Lauritzen has worldwide experience on the demolition and recycling of building materials.

INTRODUCTION

Almost all forms of human activity create one or more types of waste, which can often be recycled. This is also true of construction and demolition (C&D) waste.

After a period of time buildings are no longer suited to their original purpose and in many parts of the world natural catastrophes and wars also contribute significantly to the destruction of buildings and structures.

The reduction and utilization of waste and by-products is one of many challenges facing mankind in the modern world.

Reduction of waste in order to protect nature against pollution and in order to meet the increasing shortage of suitable dumping grounds. Utilization of waste by recycling in order to save energy and natural resources.

CONSTRUCTION AND DEMOLITION WASTE AND RUBBLE

Municipal waste consists of domestic waste, commercial waste, industrial solid waste and C&D waste.

C&D waste derives from new building and construction work and from renovation and demolition of buildings and structures. C&D waste includes among other materials concrete, asphalt and masonry, natural stone, wood, paper, plastics, cardboard, glass, metals and problem wastes. Soil is not considered a waste material, and most metals are recovered before they enter the waste stream.

C&D rubble consists primarily of broken concrete, asphalt and bricks, either in the form of separate fractions or mixed.

Close to 50 % of all municipal solid waste is C&D waste, and most of it is concrete, asphalt and masonry rubble. So in principle it should be possible to reduce by up to one half the total amount of municipal solid waste which is generated within any large city in the world by separating, processing and recycling C&D rubble apart from the main municipal solid waste stream.

In most parts of the world municipal solid waste is currently dumped at common landfills. A minor part is incinerated. Such disposal is undesirable and it is increasingly being discouraged by law and by high landfill disposal fees. This is why it would have far-reaching implications if the amounts of municipal waste could be substantially reduced by recycling C&D rubble. That is currently being done in Copenhagen, the capital of Denmark, a city of 1.5 million people (1).

RECYCLING CLEAN FRACTIONS OF C&D WASTE

It is the aim of any recycling effort to reuse waste products for the highest possible value.

Ideally old concrete should be reused as aggregate for production of new concrete. Old asphalt should be reused in new asphalt pavements, and old bricks and roofing tiles should be cleaned and reused for their original purpose.

In actual practice much old clean concrete of high quality from highway pavements and other major civil engineering structures is crushed, screened and reused as aggregate for concrete in new construction works. This has been done by the U.S. Army Corps of Engineers and numerous State Highway departments in the U.S. for at least 30 years[1] (1)(2). In order to facilitate such reuse ASTM C-33, Standard Specifications for Concrete Aggregates, allows the use of clean crushed hydraulic cement concrete as coarse aggregates on a par with natural materials (2).

Much old clean asphalt is also recycled in normal asphalt production as a matter of routine. Old bricks and roofing tiles are rarely used for their original purpose except in restoration work or for decoration.

RECYCLING MIXED C&D RUBBLE

It is the recycling of <u>mixed</u> C&D rubble that gives rise to problems. Whole bricks and roofing tiles are rarely reused for their original purpose, because it is difficult and costly to remove old cement mortar from masonry and to grade the bricks. About 50 % of all old asphalt remains as waste, which for practical reasons cannot be reused in conventional asphalt surfaces. Old concrete from demolished buildings and structures is often of so questionable a purity and quality that it is unsuitable as aggregate in the production of new concrete. These are the reasons why huge quantities of old concrete, asphalt and masonry enter the municipal solid waste stream, most of it in the form of mixed rubble.

---

[1] Rilem Report 6, Recycling of Demolished Concrete and Masonry, pp. 111-112, on recycling of pavement in USA

Practical and economic considerations suggest that optimum value reuse of concrete, asphalt and brick rubble is achieved when the rubble is crushed and used as road base or sub-base materials (1).

If all the C&D rubble generated within an average city were to be processed and used for road construction and maintenance purposes, processed C&D rubble could substitute 20 % of the consumption of natural sand, gravel and crushed stone in the city (1). For that reason it is evident that there is a potential market for processed C&D rubble provided the recycled materials can be made available at the right price and quality. As an added advantage to recycling its C&D rubble a city may be spared millions of tonnes/km of transportation of materials through the application of proper logistics (1).

PREREQUISITES FOR THE SUCCESSFUL RECYCLING OF C&D RUBBLE

Systematic recycling of C&D rubble is seldom the outcome of spontaneous development. Initially it requires the introduction of a reliable registration and collection system for C&D wastes. New laws and regulations must also be introduced and enforced.

However, first and foremost three practical requirements must be satisfied before any recycling scheme can be brought to function:

1. Selective demolition must be made mandatory.

2. Processing plants and depots for recycled materials must be established.

3. Recycled products must be produced and marketed to compete with comparable new raw materials in terms of price and quality.

Re 1. In order to achieve an acceptable economy for any recycling scheme for C&D waste, selective demolition and sorting of waste at source must be made mandatory. Selective demolition is defined as a process in which demolition takes place as 'reverse construction'. Different types and fractions of materials are removed from condemned structures and sorted at the demolition sites, so that mixing of wastes is avoided. Selective demolition and sorting of waste is now required at all major demolition sites in Copenhagen. In the process serious attempts are made to separate, store and transport concrete, asphalt and masonry rubble in three <u>clean</u> fractions, but this is not always possible. Mixed C&D rubble remains a problem.

Re 2. It is necessary to establish processing plants and depots at strategic locations within a city in order to achieve the flexibility and economy necessary for employing recycled materials. The right materials must be available in the right quantities, at the right place, at the right price and at the right time. These conditions are met by a single, large and conveniently located C&D waste recycling facility in Copenhagen (1).

Re 3. Most recycling efforts depend on subsidies in order to achieve a balanced economy. This is also true for what concerns recycling of C&D waste. The cost of

Recycling Concrete and Other Materials

selecting, sorting and processing of waste is usually higher than the value of comparable new raw materials. However, recycling of C&D waste can be made economical by imposing a fee on C&D waste, which is dumped at landfills or incinerated. The disposal fee is waived when sorted C&D waste is brought to recycling plants. From there on market economy takes over. As disposal fees gradually continue to climb, which is in the nature of all taxes and fees, a level is soon reached where commercial recycling of C&D rubble more than pays for itself (1).

Regarding product quality it is a prerequisite that recycled materials either satisfy the requirements of already existing specifications for primary raw materials or that special norms and standards be developed for recycled materials.

Danish specifications have been issued for the use of clean crushed asphalt and brick rubble as unbound road base materials (3)(4), and specifications for the use of crushed concrete are being developed. So the user is guided by official documents when he applies <u>clean</u> fractions of rubble, which are represented by the three corners of the ternary diagram, ABC, in Figure 1.

BLENDED ROAD BASE MATERIALS

In spite of all good intentions associated with selective demolition it has proved difficult and expensive to keep concrete, asphalt and masonry rubble separate in <u>clean</u> fractions on busy demolition sites. More often than not, asphalt and masonry rubble are mixed with concrete when received at recycling plants, and once the materials are mixed, they are impossible to separate again.

As a consequence of such practical difficulties the Copenhagen Recycling Centre launched two new blended products for road base construction purposes.

One is a blend of nominally 50 % crushed clean concrete and 50 % crushed asphalt. The other is a blend of nominally 55 % crushed bricks and 45 % crushed concrete. Points D and E in Figure 1 represent the two blends.

Blends of bricks and concrete are covered by Danish specifications (3), but blends of asphalt and concrete are not currently covered.

For a number of reasons the blended materials have quickly become popular among contractors. They now make up more than 90 % of all sales of recycled products from the Copenhagen Recycling Centre and there have been no complaints. This indicates a high level of customer satisfaction.

In the case of blends of crushed concrete and crushed asphalt, lubricating properties of the bitumen reduce the effort required to compact the material by rolling, and under traffic loads the mixed granular material is stronger than crushed concrete, probably due to a cohesive effect of the bitumen. Moreover, the blend supports prolonged static loads better than pure asphalt because of a proportionally lower content of bitumen. Evidently blended crushed asphalt and concrete performs better than pure crushed asphalt or pure crushed concrete. This is interesting considering that Danish specifications allow the use of pure crushed asphalt as base material for heavily trafficked roads.

Pure crushed bricks and tiles have never been popular in Denmark as road construction materials. But blends of crushed bricks and crushed concrete were well received by customers as base materials for lightly trafficked roads. The blend has good strength, it can be used under very wet conditions and it offers good support for provisional access roads. Evidently, a blend of crushed bricks and concrete has technical advantages over pure crushed bricks.

From this point of view it does not seem logical to specify the use of pure fractions of crushed concrete, asphalt or bricks, and to pay extra costs associated with the sorting, processing and storage of separate pure fractions, where mixed rubble will serve the purpose equally well or better.

As a quality measure of road base materials, "relative customer satisfaction" has been selected as an all-embracing technical and economic index of product quality.

Figure 2 is a three-dimensional perspective representation of the ternary diagram in Figure 1. On the basis of practical user experience, relative customer satisfaction is plotted along the ordinates in Figure 2 at points A, B, C, D and E. In that way points A', B', C', D' and E' are obtained. By connecting the five points by straight lines a surface emerges in space, shown hatched in Figure 2. It is suggested that this surface, based on our current knowledge, is the best estimate of the quality of the entire spectrum of blended road base materials which can be produced from crushed concrete, asphalt and bricks.

It will be seen from Figure 2 that all blended materials perform better than pure crushed bricks, which constitute the lowest quality product in our spectrum. It also appears that many binary and ternary blends can be expected to perform better than pure fractions of crushed concrete or crushed asphalt. This is not surprising. It is well known that the blending of two or more components often provides composite materials with better properties than either of the components. This is true for blends of natural sand and gravel, so why not for granular wastes? The presence of bitumen, even in small quantities, appears to impart new and useful properties to blends of mineral particles. It apparently changes the properties of the blends from predominantly brittle to more ductile.

If one is prepared to accept the fact that some blends of crushed concrete, asphalt and bricks perform better than pure materials, then it suggests itself to undertake systematic studies of key technical properties of blended C&D rubble pertaining to road construction.

Recycling Concrete and Other Materials

Properties such as compactability, bearing capacity, durability and environmental acceptability should be studied as functions of mix proportions of crushed concrete, crushed asphalt and crushed bricks.

Experimental data should be obtained in order to determine the widest possible limits within which mixes of crushed C&D rubble can be used as road construction materials in different climates and under different traffic conditions. The data should allow specifications to be prepared which will permit and promote use of blended or mixed C&D rubble for road construction purposes.

The incentive to carry out the study is considerable, because mixed rubble is simpler to handle and it is less expensive to produce than pure waste fractions.

Finally we refer to a study of the recycling of airport structures at Fornebu old airport, Oslo (5), which has demonstrated very positive results on the recycling of blended materials.

GLOBAL PERSPECTIVES OF USING MIXED C&D RUBBLE AS ROAD CONSTRUCTION MATERIALS

It is evident that large environmental and economic benefits would be associated with simple and cost-efficient handling and use of C&D rubble in quantities which may well amount to one half of all the world's municipal solid waste.

The authors have analysed the prospects of systematic recycling of C&D rubble in various cities around the world. The results are presented in Table 1. In order of importance, the three main factors affecting the prospects of recycling of C&D waste are population size and density, occurrence of, and access to natural raw materials, and level of industrialization. This is highlighted below by three examples drawn from Table 1. Other important but less predictable factors affecting the prospects of recycling of C&D rubble are wars and natural disasters such as earthquakes, hurricanes and floods.

Amsterdam in the Netherlands is the first example. The city is densely populated. Natural raw materials are scarce and the level of industrialization is high. It is not surprising that the highest possible value recycling of C&D waste is practised in the Netherlands.

Copenhagen in Denmark is our second example. The city is densely populated and the level of industrialization is high. However, because there is an ample supply of natural raw materials, economics and common sense dictate lower value recycling than in Amsterdam. Production of pure fractions of C&D rubble has proved uneconomical and the commercial market has shown a preference for blended rubble. This is in spite of the fact that public authorities for environmental reasons have discouraged the use of mixed materials.

Dacca in Bangladesh is the third example. The city is densely populated. Access to natural raw materials is difficult, and like other Third World cities Dacca struggles with the problem of reducing huge quantities of municipal solid waste that daily adds to already immense and unsanitary landfills. It would appear that the use of C&D rubble as road construction materials could substantially assist in solving both a waste disposal problem and a raw materials supply problem. This is true also for other large cities in the Third World that, contrary to Dacca, are located close to abundant raw materials deposits, but where permanently congested traffic conditions make access to the deposits difficult. Mexico City is an example in point.

However, selective demolition and sorting of waste on site cannot be made as effective in emerging as in developed nations. Mixed rubble of highly variable quality will have to be accepted if the idea is going to be useful. Practical experiences from Denmark indicate that crushed concrete rubble can be successfully used for road construction purposes even if neither the sorting out of soil, wood and plastics from the rubble nor the crushing of the rubble is strictly controlled during production of the recycled granular material. It is expected that results of the suggested systematic study of the properties of mixed C&D rubble will confirm this conclusion. What will be needed then is the preparation of more practical basic documents for simple road construction in emerging nations, including less complex code requirements, which will facilitate and promote the use of recycled waste materials.

A special Danish recycling effort has been directed towards international disaster relief and assistance in demolition of damaged buildings and recycling of building materials after catastrophes of nature and wars. Examples described in [1] are the reconstruction of war-damaged areas in Beirut in Lebanon and Mortar in Bosnia.

SUMMARY AND CONCLUSIONS

Municipal solid waste consists of domestic waste, commercial solid waste, industrial solid waste and construction and demolition waste (6).

C&D waste derives from new building and construction work and from renovation and demolition of buildings and structures. C&D rubble consists primarily of broken concrete, asphalt and bricks.

Close to 50 % of all municipal solid waste is C&D waste, and most C&D waste consists of rubble, much of it is demolished concrete. So in principle it should be possible to reduce by up to one half the total amount of municipal solid waste, which is generated within an average city by separating, processing and recycling C&D rubble apart from the main municipal solid waste stream.

In most parts of the world municipal solid waste is currently dumped at common landfills. A minor part is incinerated. Such disposal is undesirable and it is increasingly being discouraged by law and by high landfill disposal fees. This is why it has far-reaching implications if the overall quantities of municipal waste can be substantially reduced by recycling of C&D rubble. That is currently being done in Copenhagen, the capital of Denmark, a city of 1.5 million people.

Practical and economic experiences from Denmark suggest that optimum value reuse of concrete, asphalt and masonry rubble is achieved when the rubble is crushed and used as unbound road base and sub-base materials. When used for such purposes C&D rubble can substitute for up to 20 % of the consumption of natural sand, gravel and crushed stone in an average city, thereby saving natural resources.

When crushed and screened, mixed C&D rubble often has better base and sub-base making properties than pure waste fractions. Thus blends of crushed concrete and crushed asphalt have better properties than pure crushed concrete and pure crushed asphalt, just like blends of crushed concrete and crushed bricks have better properties than pure crushed bricks.

It is suggested that systematic studies of key technical and environmental properties of blended C&D rubble pertaining to road construction should be made in order to prepare specifications which will permit and promote use of the widest possible range of blended C&D rubbles. There is a strong incentive to carry out such studies because mixed rubble is simpler to handle and less expensive to produce than pure waste fractions. It is evident that large environmental and economic benefits could be associated with simple and cost-efficient handling and use of C&D rubble in quantities which may well amount to one half of all the world's municipal solid waste. The authors argue that the benefits would be equally important in emerging and industrialized nations.

Finally it is recommended that better organized ways of recycling rubble after natural disasters and wars should be developed.

REFERENCES

1. Lauritzen, E.K. and Hansen T.C., 1997, "Recycling of Construction and Demolition Waste". Danish Environmental Protection Agency. Environmental Review, 1997.

2. Hansen, T. C. (ed.), 1992, "Recycling of Demolished Concrete and Masonry" in RILEM Report 6, E&FN Spon.

3. Berg, F., Milvang-Jensen, O. and Moltved, N., 1994, "Crushed Asphalt as Unbound Roadbase". Ministry of the Environment. Danish Environmental Protection Agency. Danish Road Institute. Report 75, 1994 (in English).

4. Berg, F., Milvang-Jensen, O. and Moltved, N., 1993, "Crushed Bricks as Unbound Roadbase". Ministry of the Environment. Danish Environmental Protection Agency. Danish Road Institute. Report 71, 1993 (in Danish).

5. Torring. Mats, 2001, "Management of Concrete Demolition Waste", Ph.D. (Eng.) thesis, Norwegian University of Science and Technology.

6. WHO: "Urban Solid Waste Management". First Edition 1991-1993. IRIS, Via Campiglia no, 90. 50015 Grassina, Florence, Italy.

Table 1. Summary of the main factors affecting the prospects for successful recycling of C&D waste. Rating from "+" (doubtful) to "++++" (very profitable)

Area Code	Population Density	Natural Raw Materials Deposits	Level of Industrialisation	Example	Prospects of Successful Recycling of C&D Waste
I	High	Adequate	High	Many cities in EU, USA and the Far East, e.g. Hong Kong (China), Copenhagen (Denmark)	+++
II	High	Adequate	Low	Many megacities in the Third World, e.g. Mexico City (Mexico), Jakarta (Indonesia)	+++
III	High	Scarce	High	Amsterdam (The Netherlands)	++++
IV	High	Scarce	Low	Dacca (Bangladesh), Calcutta (India), Shanghai (China)	++++
V	Low	Adequate	High	Rural Scandinavia	++
VI	Low	Adequate	Low	Rainforests and mountain regions in developing countries	+
VII	Low	Scarce	High	Kuwait Abadan/Khorramshahr (Iran)	+++
VIII	Low	Scarce	Low	Steppelands or sandy deserts or tundras in developing countries	+

Recycling Concrete and Other Materials 45

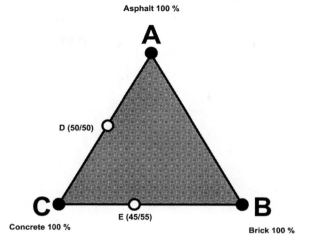

Figure 1. Ternary diagram for blends of concrete, asphalt and brick rubble

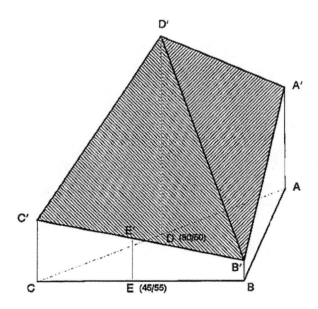

Figure 2. Three-dimensional representation of the ternary diagram in Figure 1 showing relative customer satisfaction achieved using blended rubble as sub-base in pavement construction along the ordinates. A: Asphalt 100%, B: Brick 100%, C: Concrete 100%.

SP-219—4

Guidance for Recycled Concrete Aggregate Use in the Highway Environment

by J. S. Melton

Synopsis: Recycled concrete aggregate (RCA), when used appropriately, is an excellent substitute for natural aggregates in highway construction. RCA has been used successfully in unbound applications such as base course and fill, and in bound applications as aggregate in new concrete. However, a significant amount of concrete debris is still disposed of in landfills. Barriers to concrete recycling include regulatory and policy issues, economic disincentives, environmental concerns and technical questions. This paper reviews current obstacles to concrete recycling and discusses recent developments and research that will help overcome these barriers.

Keywords: concrete recycling; demolition debris; guidance; recycled concrete aggregate; recycling; specification

Melton

Jeffrey S. Melton is the Outreach Director at the Recycled Materials Resource Center (RMRC), a Federal Highway Administration partner research center dedicated to overcoming barriers to the appropriate use of recycled materials in the highway environment. The RMRC is part of the Environmental Research Group at the University of New Hampshire.

INTRODUCTION

There is a significant amount of Portland Cement Concrete (PCC) infrastructure in the United States that is reaching the end of its working lifetime and must be replaced. This is true for the entire spectrum of concrete construction, from pavements to bridges to buildings. Looking at construction and demolition (C&D) debris as a whole, Torrig and Lauritzen (1) estimate that 400 million tons of stone, brick and concrete are generated annually worldwide. Assuming that 75% of the debris is concrete, 300 million tons of concrete debris must be processed each year. In the past, much of this material was simply thrown away. However, diminishing landfill space and increased regulation are making it increasingly expensive and difficult to dispose of concrete debris. Consider that Massachusetts is trying to reduce C&D disposal by 88% by 2010, and one of the methods is a ban on all unsorted construction materials from landfills starting in 2003 (2). The combination of political, economic and environmental pressures has many people now viewing concrete debris as a potentially valuable construction material that should be reused rather than disposed of as solid waste.

Concrete debris is typically reclaimed as recycled concrete aggregate or RCA. In this paper, RCA is broadly defined as crushed standard Portland cement concrete containing less than 5% deleterious materials by mass and containing no hazardous materials. The actual material specifications for RCA will vary depending on the application. RCA used in unbound applications will have different requirements for gradation and different limits on deleterious materials than RCA used in bound applications such as new concrete. Requirements for RCA will be discussed further in the New Guidance and Technology section.

Intuitively, recycling concrete makes sense because a tremendous amount of energy went into creating the cement and gathering the aggregate that was used to make the concrete, so it is reasonable to get as much value out of the material as possible through secondary use applications. And by recycling concrete, valuable landfill space and natural resources are preserved. This concept is part of the larger "sustainable construction" philosophy that seeks to minimize waste generation and encourage recycling in order to prevent adverse long-term environmental effects. This philosophy is becoming popular in transportation circles as well, though recycling of transportation construction materials has been practiced for decades. It was observed that highway construction consumes enormous amounts of aggregate, and that these projects could readily incorporate the large volumes of waste material being generated. Unfortunately, overzealous and inappropriate recycling of some materials has lead to the derisive term "linear landfills" to describe recycling efforts in the highway environment. However, RCA has been used successfully as base material in road construction, as well as fill for embankments. RCA

Recycling Concrete and Other Materials 49

is also increasingly used as an aggregate in new concrete construction. In both applications, RCA has been found to perform as well or better than natural aggregate (3,4,5,6,7). While such use of RCA is encouraging, a significant amount of concrete is still going into landfills or is being stockpiled.

In response to continued disposal of RCA and other materials, the Recycled Materials Resource Center (RMRC) was established in partnership with the Federal Highway Administration (FHWA) to promote the appropriate use of recycled materials in the highway environment. The RMRC has developed an active research arm to support its mission, and a significant amount of research has focused on reusing RCA. This paper discusses current barriers to concrete recycling, identifies recent guidance for using RCA in highway applications and presents results of RMRC sponsored research.

BARRIERS TO CONCRETE RECYCLING IN THE HIGHWAY ENVIRONMENT

In this paper, a barrier is defined as a practice that prevents the use of RCA in an application for which the RCA is mechanically and environmentally appropriate. A given barrier may be based on a real or a perceived issue. For instance, an end user may be resistant to using RCA because they cannot guarantee a steady supply of aggregate in the quantities required for a given project. This is a barrier based on an actual logistical problem. On the other hand, an end user may resist using RCA because 10 years ago they had difficulty obtaining sufficient quantities of RCA and they assume that the situation still exists, when the actual aggregate supply is plentiful. This is a barrier based on a perceived problem that exists due to faulty information.

The above example simplifies a very complex issue by assuming that a given barrier has a single cause, which is generally not the case. In the real world, barriers to recycling are often interconnected, which means that efforts to increase recycling must take a variety of different approaches. For the sake of discussion, different types of barriers are broken down into the following five categories:

- Regulatory and Policy Issues
- Economic Factors
- Environmental Concerns
- Technical Concerns
- Public Opinion

The list is not necessarily in order of importance, but experience suggests that public opinion must be addressed through education and outreach programs, which are generally based on information gathered while addressing the first four barriers. The following sections explore the above barrier categories more fully.

Regulatory and policy issues

Local, state and national governments can create barriers to concrete recycling by failing to provide long-term strategies for recycling and by failing to adhere to existing policies. To be successful, recycling programs need a long-term commitment from authorities at all levels of government. Simply mandating the use of RCA is not a sufficient solution because there needs to be guidance on how to implement such a program. Without guidance, more barriers may be created than removed. For example, one problem that occurs is that RCA is not considered for a project until well into the design phase, at which point it is too late to make any allowances for the differences between RCA and natural aggregate. The RCA is then rejected in favor of the natural aggregate. Had authorities insisted that the RCA be considered from the very beginning, an appropriate engineering design could have been devised.

In the highway environment, regulations and guidance come from FWHA and State Departments of Transportation (DOTs). In 1996 Chini *et al.* (8) sent surveys to all state DOTs and Puerto Rico to determine the use of RCA. They found that 32 of the 40 responding DOTs allowed the use of RAC as coarse aggregate, but only 17 of those 32 actually had RCA specifications. The most common reasons for not using RCA was a lack PCC pavement for recycling and an abundance of natural aggregate. The combination of plenty of natural aggregate and lack of specifications is a formidable barrier.

Agencies that do not have RCA specifications will often treat RCA like a traditional aggregate and apply normal aggregate specifications to its use. This can be another challenging barrier because the specifications may have been written with specific local aggregates in mind and RCA does not meet those specifications, even though it is actually an appropriate engineering material. Narrow requirements for gradation, angularity and porosity effectively act as barriers to RCA use. Fortunately, the American Association of State Highway and Transportation Officials (AASHTO) recently passed a national specification for the use of RCA in unbound applications and is considering a specification for bound applications. The implications of the specifications will be discussed in a later section.

Economic Factors

Economic reasons to recycle are based on the cost of using RCA versus using a natural aggregate. The major challenge is how to quantify those costs. Recycled concrete aggregate has costs due to demolition, crushing, sorting (including removing metal and deleterious materials), transportation and handling, as well as the avoided costs of tipping fees. Similarly, natural aggregates have costs due to mining, sorting, transportation and handling. Costs from environmental loads such as CO_2 and NO_x emissions from cement kilns, vehicles and other sources must be calculated as well, as do potential loads from contaminant leaching. CO_2 emissions have gained added importance with the signing of Kyoto Protocol to reduce green house gases. Long term performance and associated maintenance costs must also be factored in. Lastly, stability of supply for both RCA and natural aggregates must be included.

Based on the above points, Torrig and Lauritzen (1) are correct in predicating that current trends favor the economic outlook for RCA. The basic ideas are:

- Depletion of local natural aggregate sources will drive up the cost of aggregate prices due to decreased supply and to increased environmental loads and costs associated with longer transportation distances.
- Closure of existing landfills and a lack of space for new landfills will drive up tipping fees, which will encourage recycling of concrete rather than disposal.
- Improvements in demolition, processing and handling technologies will help decrease the cost of RCA.

As noted in the previous section, abundant natural aggregate is a primary reason for not using RCA. As aggregate sources decrease, both economic and policy pressures should encourage recycling concrete.

A recent survey of municipal landfill use supports the supposition that there are areas with very limited remaining landfill capacity coupled with high tipping fees. Table 1 lists states with high average tipping fees per ton and their remaining capacity in years (9). New England in particular has high average tipping fees ($75/ton in Vermont) and low remaining capacity, which makes sense given its relatively high population density. As tipping fees creep up, the economic incentive to use RCA will also increase. However, Table 2 lists states with low average tipping fees (9). While the national trend may be toward higher fees, the economic incentive from the fees must be considered on a local or regional level. In some areas, low tipping fees will hurt recycling efforts because of an economic disincentive. In such cases, the economic benefits must be considered with a long-term view.

Showing that the substitution of RCA for a natural aggregate is a viable option at the project level is another problem altogether. While it may be easy to show that RCA is environmentally and mechanically acceptable, it may be much more difficult to show that RCA is cost effective over the life of the project. Project managers do not always have the time, resources and information needed to develop a comprehensive comparison of RCA and natural aggregate, at which point the tendency is to choose the familiar material rather than take a risk with RCA. What is needed is a comprehensive set of environmentally enhanced life cycle cost analysis (LCCA) tools that take into account all of the factors described above. While some enhanced LCCA or life cycle assessment (LCA) tools exist (10,11), none currently address all aspects of the working lifetime of highway projects using RCA.

Environmental Concerns

Two environmental issues that must be addressed when using RCA in the highway environment are high pH runoff and contaminant leaching due to water flowing through RCA in unbound applications. When groundwater or rainwater moves through a layer of RCA, its pH may rise to 12 or more before leaving the layer as runoff (12). The runoff then adversely impacts the surrounding environment, particularly wetlands, by altering pH levels. The long-term runoff pH is a function of the degree of carbonation that takes place in the RCA due to CO_2. This problem has also occurred when using blast furnace slag in unbound applications due to the free lime in the slag. At one site in Maryland, the runoff pH exceeded 12.5 and was classified as a hazardous waste material. The runoff had to be collected and treated at a cost of $7,000 a week (13). To avoid this sort of problem, RCA use must be carefully considered in areas where significant water intrusion is expected. An additional complication is the formation of tufa (calcium carbonate) due to the reaction between calcium hydroxide and calcium salts in the water and atmospheric CO_2. Tufa formation is more of a technical issue that will be discussed in the next section.

Leaching of contaminants also occurs because of water intrusion into the RCA layer. Leaching can be a problem even when using clean recycled PCC pavements as a source of RCA because the cement generally contains heavy metals, though the amounts will vary from source to source. Recent work by Nilsson *et al.* (12) in Sweden has shown that Cr, Ni and Pb may exceed allowable limits, but that the leaching behavior also depends on pH and the degree of carbonation. Increased leaching of contaminants due to oil, PCBs and lead from paint may be increased when using C&D debris as a source of RCA because of detrimental materials.

Technical Issues

Even though there is a growing body of technical data and experience to support the use of RCA in highway construction, some issues persist. In the survey by Chini *et al.* (8) it was found that several states used the same material specifications for natural aggregates and RCA. While quality RCA is an excellent aggregate, it is physically different than natural aggregates due to the cement paste fraction, which can be as high as 65% of the RCA volume in the fine aggregate fraction (14). The paste may cause misleading results when the RCA is characterized using traditional protocols such as sulfate soundness and LA abrasion testing. The paste fraction may also cause the aggregate to behave differently during construction. For instance, the RCA generally has higher water absorption than natural aggregate, which must be addressed when compacting the material in unbound applications, or when used as an aggregate in new concrete. The angular nature of the aggregate means that more compactive effort may be needed to reach the required density compared to traditional aggregates. These differences, combined with a lack of guidance and specifications, have made some end users reluctant to risk using RCA because they lack sufficient experience with the material.

The high pH runoff also leads to technical problems. For instance, metal culverts may be sensitive to the runoff and may develop performance problems. Tufa formation can also occur under such conditions. Tufa will show up as a white mineral stain on the ground where runoff breaks through. The stain is unsightly and also an indicator of potential environmental problems. A less visible problem is when Tufa precipitates out of the runoff and accumulates in geotextile filter drains. The drain can be clogged by Tufa, leading to expensive repair operations.

Recycling Concrete and Other Materials 53

Another potential problem is alkali-silica reaction (ASR) in new concrete that incorporates RCA. ASR is a chemical reaction between alkaline constituents of the cement and silica in the aggregate that produces an expansive gel product that causes cracking and overall deterioration. While ASR occurs with some natural aggregates, the question here is whether or not the use of a particular batch of RCA will lead to ASR. The issue is complicated because RCA with ASR can be used in new concrete and it will not show any further signs of distress. On the other hand, if RCA with no visible ASR is used in concrete with high alkali levels, the RCA may quickly develop ASR. While examples like this are fairly rare, ASR must still be considered because it can lead to a significantly shortened working lifetime for the concrete and expensive repairs if it does occur.

Public Sentiment

Public opinion or "Attitude" is crucial to the success of concrete recycling efforts, yet it does not always get sufficient attention. In this sense, public opinion includes the opinions and attitudes of contractors, government officials and the public as a whole. Torrig and Lauritzen (1) point to Hong Kong as an example where many conditions favor recycling RCA, but public values are a main factor in preventing RCA reuse. Even when the public supports recycling in general, there is also the NIMBY or Not In My Back Yard effect to contend with, in which the public opposes local recycling projects. From experience, the reasons given when rejecting recycling include:

- Disposal in landfills is an acceptable alternative.
- Concerns about perceived risks to the environment or to public health.
- Skepticism about the quality of RCA compared to natural aggregates.
- Fears recycling will hurt local aggregate producers.

As discussed earlier, the issues can be resolved through outreach and education programs as well leadership by local authorities. In fact, these reasons should look familiar because they exist as barriers to recycling in their own right, and overcoming these barriers provides the information required to address public opinion.

NEW GUIDANCE AND TECHNOLOGY

Guidance from FHWA and AASHTO

In Spring 2002 the FHWA released a policy memorandum stating its position on recycled materials (15). The memorandum notes that the FHWA has the mission of maintaining and improving the 160,000 miles (257,440 km) of highway pavements that make up the National Highway System (NHS), and that reusing the materials used to build the original highways system "...makes sound economic, environmental and engineering sense." It also emphasizes that recycled materials must be used in an appropriate manner that "...shall not affect the performance, safety or the environment of the highway system."

The formal FHWA policy is:

1. Recycling and reuse can offer engineering, economic and environmental benefits.
2. Recycled materials should get first consideration in materials selection.
3. Determination of the use of recycled materials should include an initial review of engineering and environmental suitability.
4. An assessment of economic benefits should follow in the selection process.
5. Restrictions that prohibit the use of recycled materials without technical basis should be removed from specifications.

The memorandum is important on several levels. First, it provides support and can serve as a template for agencies at all levels of government that desire to encourage recycling. Second, there are two strong statements specifically aimed at removing barriers to the use of RCA and other recycled materials. Item 2 means that RCA should be considered before virgin aggregate and before designs are finalized. Only after RCA has been shown to be inappropriate for a specific application should virgin aggregates be used. This is a change in perspective from "Why recycle RCA?" to "Why not recycle RCA?" which provides strong guidance to the highway community. Item 5 encourages authorities to review their specifications and remove barriers to recycling RCA. This is also important because some specifications are decades old and do not reflect changes in processing and construction technologies which make using RCA much easier and effective.

The memorandum ends by encouraging cooperation between the FHWA Recycling Team, the RMRC and the AASHTO Subcommittee on Materials (SOM). The SOM passed a resolution in August 2001 that strongly supports the use of recycled materials wherever appropriate (16). As the statement noted, AASHTO member agencies have different policies on recycled materials that vary from extensive use to very little use. As such, the main thrust of the resolution was to further encourage the use of recycled materials and to promote communication of ideas and experience between its members.

AASHTO Specification M-319-02

The recommendation to AASHTO and supporting white paper (17) that became AASHTO Standard Specification for Reclaimed Concrete Aggregate for Unbound Soil-Aggregate Base Course [M 319-02] (18) was developed by Chesner Engineering in Commack, NY with funding from the RMRC. An advisory group composed of members from the DOTs of California, Florida, Illinois, Massachusetts, Michigan, Minnesota, New Hampshire, New Jersey, New York State, North Carolina, Ohio, Pennsylvania, Texas and Wisconsin provided guidance and input for the draft. The goal was to develop a specification that combined the experiences of various DOTs into a guidance document that would encourage RCA recycling.

The specification addresses several of the questions raised in the technical issues section. Note 1 and Appendix A discuss the increased water absorption by RCA and how the moisture-density curves shift for RCA so that the optimum moisture content is generally higher than for natural aggregates. Appendix A gives detailed guidance on quality control measures and how to compact the RCA in a base course. Notes 2 and 3, with Appendix B, address the issue of high pH runoff and tufa formation. Note 6, Section 7 and Appendix D give guidance for dealing with contaminated waste and deleterious materials in the RCA.

Note 5 and Appendix C address the issue of soundness testing for RCA. Several State DOTs require sulfate soundness testing when characterizing an aggregate. However the sulfate solution can actively attack the cement paste, leading to premature determination of the RCA and correspondingly high loss values. The question then becomes whether or not to use sulfate soundness testing, and if not, what other test should be used? The specification gives different options for alternative soundness testing procedures as well as a "No-Test" alternative.

Lastly, Note 4 reminds the engineer that over time an RCA base course may self-cement, leading to an increase in strength and stiffness as well as a corresponding decrease in permeability. The added strength and stiffness is a benefit because it provides better support for the surface layer and may reduce rutting. However, if the base course is supposed to act as a drainage layer, the reduction in permeability may offset the benefits of increased strength. The specification suggests removing the fine portion of the RCA if self-cementing will cause adverse consequences.

The specification does not address ASR because it is not an issue in unbound applications. ASR may be a problem in new concrete construction with reactive aggregate because the expansive gel causes cracking and may accelerate freeze-thaw damage due to increased water penetration. As mentioned, there is a draft specification under consideration by the AASHTO Subcommittee on Materials for RCA in concrete that will address ASR (19).

Alkali-Silica Reaction

The RMRC has sponsored research by Dr. David Gress of University of New Hampshire to further investigate ASR in concrete with RCA as aggregate. The main goals of the research were to develop means to quickly and accurately test for potential ASR reactivity as well as find methods to mitigate detected ASR. The reason for an accelerated ASR test is that current testing protocols can take a year or more to verify ASR activity. Project managers and contractors need to know within a week to a month whether the RCA (or natural aggregate) will react so that appropriate steps can be taken.

During the course of this research, test protocols ASTM C 1260 and C 1293 were modified for accelerated testing of RCA. ASTM C 1260 was created for accelerated ASR testing, but it is not applicable to RCA because it requires the aggregate be finely ground and cast into beams with a 25 mm cross section (20,23). Grinding would destroy the RCA and produce test results that would not be representative of field behavior where the RCA is used intact. Larger samples in the form bars with a 76 mm cross section and 286 mm length as well as 76 mm cubes with and without cast in longitudinal holes were used with RCA from a section of Interstate 95 in Maine. The increased sample size compared to ASTM C 1260 allowed the RCA to be used intact. Results suggest that the limiting expansion can be determined at 28 days compared to 120 days for the unmodified samples (24).

ASTM C 1293 protocol was modified by sealing the samples, using larger samples and by applying a dc current to increase the ASR reaction rate. The standard requires evaluation of expansion after 365 days of testing (21). The results showed that sealing the samples to maintain moisture content significantly accelerated the test from 365 days for the standard C 1293 to 200 days for sealed samples. Using cubic samples with a cast in holes further reduced the test period to 50 days (24). It was also found that water absorption by RCA contributed significantly to the measured expansion, especially during the initial period, and that the RCA should be fully saturated before mixing and testing. This research is on going because it is still unclear if the old paste in RCA re-activates and contributes alkali to the new concrete mix.

This research also showed that ASR problems in concrete with RCA could be mitigated by sufficient dosage of fly ash, ground granulated blast furnace slag (GGBFS), silica fume or lithium nitrate (22). But compared to the virgin aggregate, RCA needs higher dosages. The research considered mitigation strategies that were added to concrete during the mixing phase. Work is continuing on the lithium nitrate because there is the possibility that it can be applied in solution to treat ASR distressed concrete that is already in place.

FUTURE RESEARCH

Life Cycle Analysis

The RMRC is funding a research project directed by Arpad Horvath at the University of California at Berkeley to develop a life cycle analysis (LCA) model and a computer based decision tool to help managers and engineers compare recycled materials with natural materials for highway construction projects (25). A survey of available data shows that many build and maintenance decisions are based on initial costs and accumulated experience. Procedures for making construction and maintenance decisions also vary from state to state and from region to region. Without the environmentally enhanced life cycle cost analysis (LCCA) tools mentioned earlier, it is very difficult to include environmental loads and long-term costs for natural aggregates, let alone to compare RCA and natural materials. The goal is to create a unique model that combines traditional economics based tools with tools to quantify environmental effects as well. Managers and designers will then be able to make informed decisions about which material is most appropriate for a given situation.

Most of the research to date has been collecting cost data for construction and maintenance using both recycled and traditional materials as well as collecting data on avoided costs such as tipping fees. Costs for environmental loads are also being collected, though the data is sparse, so some costs are being estimated using LCCA models. Even without the decision tool, this database will be very useful for showing how the long-term costs of construction with RCA compares to construction with traditional materials.

SUMMARY

Recycled concrete aggregate or RCA is an excellent material for use in bound and unbound applications in the highway environment. However, there are still many barriers preventing more widespread use of RCA. This paper describes current barriers, and identifies needed guidance, information and technology needed to overcome the barriers. New guidance from FHWA and AASHTO regarding RCA and recycled materials is presented, as is recent research on RCA sponsored by the RMRC. By sharing this information between the concrete and transportation communities, it is hoped that concrete recycling will be encouraged.

ACKNOWLEDGEMENTS

This material is based on work supported by the Federal Highway Administration under Cooperative Agreement No. DTFH61-98-X00095 with the Recycled Materials Resource Center at the University of New Hampshire, Durham, New Hampshire.

REFERENCES

1. Torring, M., and E. Lauritzen, Total Recycling Opportunities-Tasting the Topics for the Conference Session. *Proceedings of the International Conference on Sustainable Concrete Construction*, Dundee, Scotland, pp. 501-510, 2002.

2. Allison, P., McQuade, J. and S. Long, Diverting C&D Debris: The Interplay of Policies and Markets, *Resource Recycling*, December 2002, pp. 1-3.

3. Arm, A., Johansson, H. G., and K. Ydrevik, Performance-Related Tests on Air-Cooled Blast-Furnace Slag and Crushed Concrete, VTI Activities in The European ALT-MAT Project, *Beneficial Use of Recycled Materials in Transportation Applications*, T. Eighmy (Ed), Arlington, Virginia, USA, November 13-15, pp. 237-248, 2001.

4. Forsman, J. and H. Höynälä, Reclaimed Crushed Concrete, *By-products and Recycled Materials in Earth Structures: Materials and Applications*, H. Mäkelä and H. Höynälä (Ed), National Technology Review, Helsinki, Finland, pp. 33-38.

5. Bennert, T., Papp, W. J., Maher, A. and N. Gucunski, Utilization of Construction and Demolition Debris Under Traffic-Type Loading in Base and Subbase Applications, *Transportation Research Record 1714*, TRB, National Research Council, Washington, D.C., 2000, pp. 33-39.

6. Chini, A. R., Kuo, S., Armaghani, J. M. and J. P. Duxbury, Test of Recycled Concrete Aggregate in Accelerated Test Track, *J. Transportation Engineering*, Vol. 127, No. 5, pp. 486-492, 2001.

7. Meinhold, U., Mellmann, G., and M. Maultzsch, Performance of High-Grade Concrete with Full Substitution of Aggregates by Recycled Concrete, *Third CANMET/ACI International Symposium on Sustainable Development of Cement and Concrete*, V. M. Malhotra (Ed), San Francisco, CA, USA, September 16-19, 2001.

8. Chini, S. A., Sergenian, T. J. and J. M. Armaghani, Use of Recycled Aggregates for Pavement, *Proceedings of the Forth Materials Engineering Conference*, Washington, DC, 1996.

9. Goldstein, N. and C. Madtes, The State of Garbage in America, *BioCycle*, Vol. 42, No. 12, p. 48, 2001.

10. Carnegie Mellon University Green Design Initiative, Economic Input-Output Life Cycle Assessment, http://www.eiolca.net/, Accessed March 28, 2003.

11. Eskola, P., Mroueh, U., and A. Nousiainen, Life-Cycle Inventory Analysis Program for Road Construction – Development and Experiences of Use, *Beneficial Use of Recycled Materials in Transportation Applications*, T. Eighmy (Ed), Arlington, Virginia, USA, November 13-15, pp. 249-258, 2001.

12. Nilsson, U., Håkansson, K., and A. Fällman, Leaching Tests for the Evaluation of Environmental Suitability of Crushed Concrete and Air-Cooled Blast Furnace Slag, *Beneficial Use of Recycled Materials in Transportation Applications*, T. Eighmy (Ed), Arlington, Virginia, USA, November 13-15, 2001.

13. Boyer, B. W., Alkaline Leachate and Calcareous Tufa Originating from Slag in a Highway Embankment near Baltimore, Maryland, *Transportation Research Board 1434*, TRB, National Research Council, Washington, D.C., pp. 3-7, 1994.

14. Commonwealth Scientific and Industrial Organization (CSIRO), Guide to the Use of Recycled Concrete and Masonry Materials, HB 155-2002, 1st Edition, Standards Australia, 2002.

15. Federal Highway Policy Memorandum. FHWA Recycled Materials Policy. http://www.fhwa.dot.gov/legsregs/directives/policy/recmatpolicy.htm, Accessed March 25, 2003.

16. AASHTO Subcommittee on Materials Recycling Policy, Resolution on Use of Recycled Materials, http://www.rmrc.unh.edu/Resources /AASHTO/AASHTOSubcommitteeMaterialsRecyclingPolicy.asp, Accessed March 25, 2003.

17. White Paper and Specification for Reclaimed Concrete Aggregate for Unbound Soil Aggregate Base Course, Recycled Materials Resource Center, http://www.rmrc.unh.edu/Research/Rprojects/Project13/Specs/docs/RCADraft.pdf, Accessed April 24, 2003

18. AASHTO, Standard Specification for Reclaimed Concrete Aggregate for Unbound Soil-Aggregate Base Course, M 319-02, *Standard Specifications for Transportation Materials and Methods of Sampling and Testing*, 22nd Edition, pp. M 319-1-M 319-8, 2002.

19. Draft White Paper for Reclaimed Concrete Aggregate for Use as Coarse Aggregate in Portland Cement Concrete, Recycled Materials Resource Center, http://www.rmrc.unh.edu/Research/Rprojects/Project13/Specs/RCPC/p13RCPC.asp, Accessed April 24, 2003

20. ASTM, Standard Test Method for Potential Alkali Reactivity of Aggregates (Mortar-Bar Method), ASTM C 1260-94, American Society for Testing and Materials, Philadelphia, PA, Vol. 04.02, Section 4, 1994.

21. ASTM, Standard Test Method for Concrete Aggregates by Determination of Length Change of Concrete Due to Alkali-Silica Reaction, ASTM C 1293-95, American Society for Testing and Materials, Philadelphia, PA, Vol. 04.02, Section 4, 1995.

22. Scott IV, H., and D. Gress, Mitigating Alkali Silicate Reaction in Recycled Concrete, *Beneficial Use of Recycled Materials in Transportation Applications*, T. Eighmy (Ed), Arlington, Virginia, USA, November 13-15, pp 959-968, 2001.

23. Gress D., and R. Kozikowski, Accelerated ASR Testing of Recycled Concrete Pavement, *Transportation Research Record 1698*, TRB, National Research Council, Washington, D.C., pp 1-8, 2000.

24. Gress, D., Kozikowski, R., and H. Scott IV, Accelerated ASR Testing of Concrete Prisms, *Beneficial Use of Recycled Materials in Transportation Applications*, T. Eighmy (Ed), Arlington, Virginia, USA, November 13-15, pp 563-574, 2001.

25. Horvath, A., Pacca, S., and C. Debaille, Framework and Tool for Environmental Life-cycle Assessment of Pavements, Technical report, University of California, Berkeley, Department of Civil and Environmental Engineering, 2002.

Table 1. States with high average tipping fees (After Goldstein and Madtes (9))

State	Average Tipping Fee ($/ton)	Remaining Capacity (Years)
Vermont	75.00	6.3
Massachusetts	67.00	<2
New Hampshire	66.00	8
Maine	65.00	12-18
Delaware	58.50	30
Hawaii	50.00	1-15

Table 2. States with low average tipping fees (After Goldstein and Madtes (9))

State	Average Tipping Fee ($/ton)	Remaining Capacity (Years)
Nevada	18.00	>50
Oklahoma	20.00	20
Mississippi	25.00	20
North Dakota	25.00	20
Oregon	25.00	40
Missouri	29.53	9

Mitigating Alkali Silica Reaction in Recycled Concrete

by H. C. Scott IV and D. L. Gress

Synopsis: This study investigated the reactivity of concrete containing recycled concrete aggregates (RCA) that had shown distress due to alkali silica reaction (ASR). The investigation evaluated several mitigation techniques to control ASR in concrete containing potentially reactive RCA. Mitigation work was done with three different aggregate types; an igneous fine-grained quartzite aggregate locally called blue rock, a non-reactive limestone, and RCA containing blue rock aggregate. These aggregates were used to investigate various mitigation techniques to prevent ASR from occurring in concrete containing RCA. The mitigation strategies include the use of class F fly ash, ground granulated blast furnace slag (GGBFS), lithium nitrate, silica fume blended cement and low alkali cement. These materials were incorporated into concrete mixes by cement substitution and direct application. These mitigation strategies showed potential in controlling ASR distress in RCA concrete. Mortar bars and concrete prisms were used to investigate the mitigation strategies by following standard and modified versions of ASTM C 1260 and ASTM C 1293 specifications to evaluate expansion caused by ASR. The modified versions of ASTM C 1260 were found effective in evaluating potential ASR expansion using conventional aggregates.

Keywords: alkali silica reaction (ASR); concrete recycling; durability; fly ash; ground granulated blast furnace slag; lithium nitrate; recycled concrete aggregate; silica fume

ACI member Hugh C. Scott IV is employed as a Civil Engineer by Appledore Engineering, Inc. He is presently completing his MS degree in Civil Engineering with a concentration in Materials at the University of New Hampshire. He is a member of the American Concrete Institute (ACI) and the American Society of Civil Engineers (ASCE).

ACI member Dr. David L. Gress is a Professor of Civil Engineering and Associate Director of the Recycled Materials Resource Center at the University of New Hampshire. He is a member of ACI 555, Recycling Concrete and Other Materials, TRB A2E01, Durability of Concrete, and TRB A2E06, Basic Research and Emerging Technologies Related to Concrete.

RESEARCH SIGNIFICANCE

Recycled concrete aggregate (RCA) has been successfully utilized in the production of portland cement concrete (PCC) for many years, however there is little information about using ASR distressed RCA in PCC. It has been shown that mitigation strategies can control ASR distress when applied to new concrete mixes in the correct dosages (1). Class F fly ash, GGBFS, lithium nitrate, silica fume blended cement and low alkali cement are all proven mitigation strategies used to control ASR (1). Information on these mitigation strategies ability to control ASR distress with ASR distressed RCA has been limited and inconclusive. This topic is timely due to increased alkali contents in modern day portland cements, recent interest in recycling caused by depleting natural aggregate sources and the desire to make concrete green.

INTRODUCTION

ASR distress in concrete structures was first observed in the United States in 1940 (2). Much of the concrete infrastructure in the United States is inadequate and in need of rehabilitation or reconstruction. It will cost many trillions of dollars to complete this rebuilding process. The use of reactive aggregate in concrete results in ASR distress if the alkali content is high. Although non reactive aggregates do exist, they are often located in remote locations and transportation cost prohibits their use. A viable alternative to aggregate depletion is to recycle existing concrete. This not only reduces construction costs and demolition waste, but also helps take pressure off depleting aggregate supplies.

The physical and chemical makeup of a given RCA needs to be understood before it can be used in new concrete. There are two possibilities when using RCA as aggregate in concrete. The RCA can come from concrete that has shown signs of ASR distress or concrete that has not. If concrete has shown ASR distress the reaction may be complete or partially complete. A RCA obtained from concrete that has completely reacted may never undergo ASR distress when put into new concrete, however when the RCA has ASR the new concrete may experience ASR distress. If concrete has shown no ASR distress in the field, the process of recycling and incorporating the RCA into new concrete may trigger the reaction if a more aggressive alkali environment is introduced and the aggregate in the RCA was potentially reactive. The alkali present in the reacted

ASR gel and deicing salts contained within the RCA can have a detrimental effect on new PCC. This internal alkali may have the ability to reactivate and contribute significantly to ASR distress in RCA concrete. The alkali from portland cement and RCA in concrete creates very high initial pH levels that are more aggressive than those seen in concrete where the cement is the only alkali source.

MATERIALS

The RCA source for this study was a section of Interstate-95, near Gardner, Maine, which showed distress from ASR. The concrete was made with an igneous fine-grained quartzite coarse aggregate locally called blue rock and a non reactive limestone from Vermont for a baseline reference. The fine aggregate was non reactive glacial sand from Ossipee, New Hampshire. The aggregate and cement properties are presented in Tables 1 and 2 respectively. The mitigating materials evaluated, as presented in Table 3, are class F fly ash and ground granulated blast furnace slag (GGBFS). Class F fly ash and GGBFS was substituted with the control cement at 25% and 55% respectively. Silica fume blended cement and low alkali cement were evaluated without the benefit of other mitigation strategies.

Lithium nitrate was used at a 100% dosage, based on a lithium to alkali molar ratio of 0.74. Although the dosage normally is based only on the cement, in the case of the RCA it was based on the availability of the water soluble alkali contributed from the aggregate as well as the cement. The soluble alkali content of the blue rock and RCA, as determined by the Bérubé (3,4,5) procedure, were 3.5 % and 5.15 % respectively. The lithium nitrate dosage for RCA concrete was increased to 125% to take into account alkali within the concrete system that would be attributed to the RCA. It is assumed a significant amount of the RCA alkali is related to the use of deicing salts as well as the solubility of the alkali within the ASR gel present.

The mix proportions for the original RCA concrete and the concrete test mixes are presented in Table 4. A volume of course aggregate per unit of volume of concrete factor of 0.63 was use to proportion all test mixes. The RCA aggregate source was stored in sealed containers and purged with nitrogen to minimize carbonation. All other materials were stored at room temperature in the laboratory.

METHODS

Three testing methods were used in this research to look at potential ASR distress in concrete test samples. ASTM C 1260 was used to show the effects of the mitigation strategies on the blue rock, limestone and recovered RCA blue rock aggregate (6,7). A modified ASTM C 1260 prism test was used to evaluate blue rock and RCA concrete beams (8). For comparative purposes the standard ASTM C 1293 test was also performed on all mixes (9). The ASTM C 1293 test method is highly recognized as an essential ASR test in evaluating a PCC (10). Details of these modified tests have been published in past work (11,12,13, 14, 15, 16, 17).

ASTM C 1260 Mortar-Bar

This accelerated test method indicates, within sixteen days of sample mixing, the potential for ASR distress of an aggregate (6). ASTM C 1260 is a screening test suggesting additional evaluations be determined if expansions are higher than the selected expansion limit commonly accepted to be 0.10 %.

The standard mortar-bar method was used to evaluate blue rock, limestone and recovered aggregate from RCA with all mitigation materials, except low alkali cement. The aggregate was recovered from the RCA by using alternating cycles of freezing with liquid nitrogen and thawing using a microwave oven as per the procedures of Bérubé (7). It is not possible to evaluate RCA with the standard ASTM C 1260 test because it is impossible to prepare the required grading without destroying the integrity of the paste within the RCA.

A modified version of ASTM C 1260, extending the testing period to 28 days, was used to determine if the mitigation strategies that were incorporated into mixtures through cement substitution could control ASR distress with the blue rock aggregate. These results were used to select the appropriate mitigation strategies for evaluating the RCA. Only those mitigation strategies that controlled expansions with the blue rock aggregate were used with the RCA.

When lithium nitrate samples were evaluated the sodium hydroxide soak solution was adjusted to have the same amount of lithium as the dosed amount within the sample while maintaining a 1 N NaOH solution. The addition of lithium nitrate to the sodium hydroxide soak solution was done according to published criteria from the lithium nitrate supplier for this research. This was done to prevent the very mobile small lithium ion from leaching from the samples.

Modified ASTM C 1260 Prism

This modified test was used so the actual RCA PCC mixture could be evaluated in an accelerated testing environment. This test was adopted from work done by Benoit Fournier at the International Centre for Sustainable Development of Cement and Concrete (8). The ASTM C 1293 prism is used with this testing procedure, which allows RCA concrete samples to be evaluated, as no pulverizing methods are necessary. Concrete beams are submerged in 1 normal NaOH and are considered to fail if the expansion exceeds 0.04% within 28 days. The failure criteria for prisms incorporating mitigation strategies, through cement substitution, are 0.04% within 56 days. These expansion criteria were developed on correlations of field data and ASTM C 1293 prism test results (8). This allows for a much faster result than the one to two year testing period for the ASTM C 1293 test.

ASTM C 1293 Prism

This test method evaluates a full-scale concrete mix for potential ASR distress in a one-year period (two years if mitigated with cement substitution) (9). This test also uses prisms so that regular size aggregate particles can be incorporated in the concrete test mixes. ASTM C 1293 is highly accepted as being the best ASR detection method for full-scale concrete test samples.

RESULTS AND DISCUSSION

ASTM C 1260 Mortar Bar

The results of the control cement and lithium nitrate concrete mixtures with blue rock aggregate and RCA are presented in Figure 1. The 14-day expansions of the control cement mixtures with the blue rock aggregate and RCA were well above 0.10%, suggesting mitigation is required. Lithium nitrate mitigation was very effective in reducing the expansion as expected.

The effect of class F fly ash, GGBFS and silica fume blended cement is presented in Figure 2 where the testing was continued to 28 days because the mixes were mitigated through cement substitution. The Class F fly ash and GGBFS reduced the expansion level below 0.10% at 14 and 28 days, however the silica fume did not control expansion.

These data show class F fly ash, GGBFS and lithium nitrate are useful mitigation strategies to control ASR expansions. Silica fume blended cement has in general shown effective results in controlling ASR expansion for most aggregates at a substitution higher than contained in the blended cement (1)

Modified ASTM C 1260 Prism

The modified ASTM C 1260 prism test results at 28 and 56 days were identical to the mortar bar test results for all blue rock mixtures as shown on Figures 3 and 4. A 100% lithium nitrate dosage controlled expansion below 0.04% for blue rock and RCA at 28 days as shown by Figure 3. The class F fly ash and GGBFS cement substitution mixes controlled deleterious ASR expansion below 0.04% at a 56-day test duration as shown by Figure 4. Although the 8% silica fume blended cement blue rock mix kept the expansion less than 0.04% at 28 days, it did not after 56 days. The equivalent silica fume mix using RCA did not reduce the 56-day expansion to less than 0.04%, which is consistent with what was expected.

Identical results with the ASTM C 1260 mortar-bar test suggest the modified ASTM 1260 prism test is valid for use with the virgin and RCA blue rock aggregate.

Figure 5 shows the correlation between the modified prism test at 28 days and the standard mortar bar test at 14 days. This shows the excellent correlation between the two tests by agreeing when the samples pass as well as fail the expansion criteria. Figure 6 shows a similar relationship between the two modified tests but at 56 and 28 days. These criteria of increased testing periods (56 versus 28 for the prisms and 28 versus 14 for the mortar bars) are required when cement substitution mitigation is utilized. If the testing times are not increased some mitigation procedures may not be conservative and ultimately lead to premature failure in the field.

ASTM C 1293 Prism

Blue rock mixtures showed identical results for high alkali and lithium nitrate mitigation samples when tested by ASTM C 1293 as shown by Figure 7. The lithium nitrate mix effectively controlled the one-year expansions whereas the high alkali mix did not. Although the use of ultra low alkali cement is not realistic when used as a mitigation strategy, it is extremely effective as expected. The difference between the blue rock and RCA mixes with low alkali cement is due to the existing alkali in the RCA. The high alkali mix with blue rock and RCA showed failure expansions greater than 0.04% well before the one-year testing limit. Unlike results seen with the modified ASTM C 1260 prism data, lithium nitrate was not as effective controlling expansions, however even though it was close 0.04% limit it did pass the test at 365 days. The RCA lithium nitrate failed the test after approximately 200 days of testing suggesting that a 125% dose is not sufficient for the RCA and the high alkali control cement used.

ASTM C 1293 prism cement substitution mixtures require a two-year test duration period. Figure 8 shows that although no samples have completed two years of testing the RCA mixes with class F fly ash, GGBFS and the silica fume mitigations have already exceeded the 0.04% expansion limit. These results do not compare well with the modified ASTM C 1260 prism testing data. The blue rock aggregate mitigated mixtures are expected to finish the two-year duration below the failure expansion limit.

Only low-alkali cement was able to control the expansion limit of the RCA mixture. The ASTM C 1293 prism results show comparable data to both ASTM C 1260 tests for the blue rock aggregate mixtures, but not with the blue rock RCA mixtures.

Close observation of all the RCA ASTM C 1293 test data showed very high early expansions, which were not present with the blue rock mixes. It was thought this might be related to moisture expansion within the RCA forcing the prisms to swell at early ages. In an effort to explain the early expansion, additional testing was performed with the RCA to determine the effect of increasing the initial moisture content of the RCA prior to mixing.

The effect of this swelling (wetting and drying expansion) was evaluated by either presoaking or vacuum saturating the RCA with water prior to mixing so it would expand prior to mixing and therefore prevent early expansion of the concrete. No additional water was added so as to prevent the loss of any alkali in the test prisms. The results from the saturation procedures are presented in figures 9 through 12. The initial expansions of all RCA mixtures were decreased when compared to results seen in the previous ASTM C 1293 prism data. Saturating the aggregate before putting it into the concrete mixtures prevented early swelling of the RCA mixes. Although the time of testing is not near completion, it appears the RCA mitigation mixes, except for the lithium nitrate, which appears to have flattened out, will still exceed the two year expansion limit of 0.04%. This suggests that increased dosages will be required to make the mitigation strategies viable.

The correlation between the ASTM C 1293 two year prism test and the ASTM C 1260 28 day mortar bar test is shown in Figure 14 for those mixes that did not have early age swelling. The correlation between these two tests is very encouraging for the various forms of mitigation strategies evaluated. The two testing procedures give consistent results in either accepting or failing a given mitigated mix. This not only suggests a lower cost of testing but also gives confidence in determining the equivalence of testing for two years in only one month.

CONCLUSIONS

The following conclusions can be made on mitigating alkali silicate reaction in the RCA PCC evaluated. Although these may or may not pertain to other aggregates and RCA it seems reasonable they would be similar.

1. RCA PCC is more susceptible to ASR than the original aggregate due to preexisting internal alkali.
2. The early age moisture swelling of RCA PCC prisms is the physical result of the RCA paste expanding and is not related to ASR.
3. RCA saturation premix procedures decrease early-age moisture expansion in RCA PCC.
4. Standard ASTM C 1260 mortar bar testing results are comparable with modified ASTM C 1260 concrete prisms testing results.
5. ASTM C 1293 test data compare well with the modified ASTM C 1260 prism and ASTM C 1260 mortar bar testing for conventional and RCA.
6. Presently used ASR mitigation strategies require higher doses to mitigate ASR with RCA.

ACKNOWLEDGEMENTS

The authors would like to thank the US DOT FHWA and the Recycled Materials Resource Center for funding and support on this project.

Hugh C Scott IV can be contacted by postage, email, or phone by the following: Appledore Engineering, Inc., 15 Rye Street, Suite 305, Pease International Tradeport, Portsmouth, NH 03801, hscott@appledoreeng.com, (603) 433-8818.

Dr. David L. Gress can be contacted by postage, email or phone by the following: 235B Kingsbury Hall, Civil Engineering Department, University of New Hampshire, Durham, NH 03824, david.gress@unh.edu, (603) 862-1410.

REFERENCES

(1) Malvar, L.J., G.D. Cline, D.F. Burke, R. Rollings, T.W. Sherman, and J. Greene, October 2001, "Alkali-Silica Reaction Mitigation State-Of-The-Art", Technical Report TR-2195-SHR, Naval Facilities Engineering Service Center, Port Hueneme, Ca.

(2) Mehta, P. Kumar, and Paulo J.M. Monteiro, *Concrete Structure, Properties, and Materials*, Prentice-Hall, Inc., 1993

(3) Appendix C, "Test Procedure for Determination of Water-Soluble Alkali Content", Canmet, International Workshop on Alkali-Aggregate Reactions in Concrete: Occurrences, Testing and Control, Halifax, Nova Scotia, May, 1999, pg. 289.

(4) Durand, Benoit, 2000, "A Note About Alkali Contribution From Aggregates in Concrete Affected by ASR", *Proceedings of the 11th International Conference on Alkali-Aggregate Reaction in Concrete*, June, Quebec City, QC, Canada.

(5) Bérubé, M., Pedneault, A., Frenette, J., and Rivest, M., "Laboratory Asessment of Potential for Future Expansion and Deterioration of Concrete Affected by ASR", *Proceedings CANMET/ACI International, Workshop on AAR in Concrete*, Dartmouth, Nova Scotia, October 1995, pp 267-291.

(6) ASTM C 1260-94, 2001. "Standard test method for Potential Alkali Reactivity of Aggregates (Mortar-Bar Method)", Annual Book of ASTM Standards, Section 4, Volume 4.02 (Concrete and Aggregates).

(7) Appendix B, "Test Procedure for Extraction of Aggregates from Cores", Canmet, International Workshop on Alkali-Aggregate Reactions in Concrete: Occurrences, Testing and Control, Halifax, Nova Scotia, May, 1990, pg. 289.

(8) Telephone conversation with Benoit Fournier, 6/17/02.

(9) ASTM C 1293-01, 2001. "Standard test method for Determination of Length Change of Concrete Due to Alkali-Silica Reaction", Annual Book of ASTM Standards, Section 4, Volume 4.02 (Concrete and Aggregates).

(10) Touma, Wissam E., David W. Fowler, Ramon L. Carrasquillo, Kevin J. Folliard, and Norman R. Nelson, 2001, "Characterizing Alkali-Silicate Reactivity of Aggregates Using ASTM C 1293, ASTM C 1260 and Their Modifications", Paper No. 01-3019, Transportation Research Record 1757, Journal of the Transportation Research Board, Washington, D.C.

(11) Gress D., Kozikowski, R., "Accelerated ASR Testing of Recycled Concrete Pavement," Transportation Research Record, Concrete 2000, Materials and Construction, No. 1698, 2000, pp 1-8.

(12) Gress D., Kozikowski, R., "Accelerated ASR Testing Of Concrete Prisms Incorporating Recycled Concrete Aggregates," *Proceedings, of the 11th International*

Conference on Alkali-Aggregate Reaction in Concrete, M. Bérubé, B. Fournier, and B. Durand Ed., Quebec City, Canada, 2000, pp. 1139-1147.

(13)   Gress, D., Kozikowski, R., and Eighmy, T., "Accelerated ASR Testing of Recycled Concrete", Waste Materials in Construction, Proceedings of the International Conference on the Science and Engineering of Recycling foe environmental protection, Harrogate England 31 may, 1-2 June 2000, pp 221-233.

(14)   Kozikowski, R., "Accelerated ASR Testing of Concrete Using Recycled aggregates," Thesis submitted to the University of New Hampshire, December, 2000, 1-135.

(15)   Scott IV, H., and Gress, D., "Mitigating Alkali Silicate Reaction in Recycled Concrete", Beneficial Use of Recycled Materials in Transportation Applications, Arlington Virginia, November 13-15, 2001, pp 959-968.

(16)   Gress, D., Kozikowski, R., Scott IV, H., "Accelerated ASR Testing of Concrete Prisms", Beneficial Use of Recycled Materials in Transportation Applications, Arlington Virginia, November 13-15, 2001, pp 563-574

(17)   Scott IV, H.C., "Mitigating Alkali Silicate Reaction in Recycled Concrete", Master's Thesis, University of New Hampshire, 2003.

Table 1: Aggregate Material Characteristics

Aggregate	Specific Gravity (Gs)	% Absorption	DRUM (lb/ft3)	Fineness Modulus	Moisture Content (%)
Natural Blue Rock	2.69	0.47	96.76	-	0
RCA Blue Rock	2.35	4.10	79.96	-	0
Limestone	2.74	0.50	103.4	-	0
Sand	2.75	1.13	-	2.68	0

Note: DRUW or Dry Rodded unit Weight

Table 2: Cement Material Characteristics

Material	Specific Gravity (Gs)	Na_2Oeq (%)	MgO (%)	Autoclave Expansion (%)	CaO (%)	Loss on Ignition (%)
Control Cement	3.15	1.31	3.1	0.13	61.8	1.37
Low Alkali Cement	3.15	0.26	1.1	0.09	NA	1.4
Silica Fume Cement (8%)	2.75	0.95	2.7	0.07	58.5	0.9

Table 3: Class F Fly Ash and GGBFS Material Characteristics

Material	Specific Gravity (G_s)	Na_2Oeq (%)	CaO (%)
Class F fly ash	2.22	0.81	1.46
GGBFS	2.85	0.41	N/A

Table 4: PCC Mix Proportions for Original RCA Concrete and Research Concrete

Concrete Mixture:	RCA PCC Mixture (Original)	PCC Mixture (Present Research)
Cement Content:	+/- 600 lb/yd^3	708 lb/yd^3
Water-Cement Ratio (W/C):	0.40 - 0.45	0.45
Air entrainment Content:	6 %	6 %
Maximum Aggregate Size:	1-1/2 inches	0.75 inches

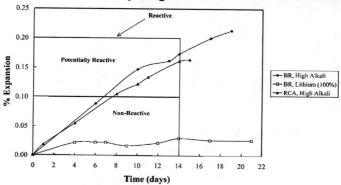

Figure 1: ASTM C 1260 Mortar-Bar (Lithium Nitrate Mitigation)

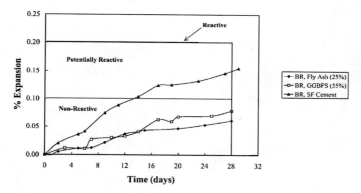

Figure 2: ASTM C 1260 Mortar-Bar (Cement Substitution Mitigation)

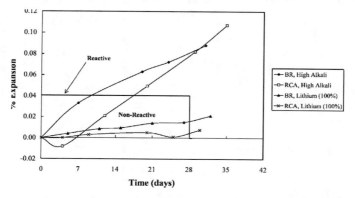

Figure 3: Modified ASTM C 1260 Prism (Lithium Nitrate Mitigation)

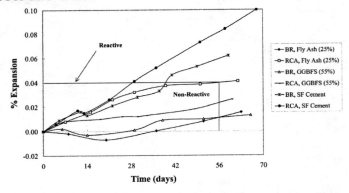

Figure 4: Modified ASTM C 1260 Prism (Cement Substitution Mitigation)

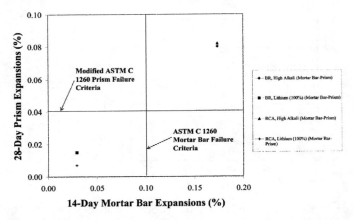

Figure 5: Modified ASTM C 1260 28-Day Prism versus ASTM 1260 14-Day Mortar Bar Results

Recycling Concrete and Other Materials 73

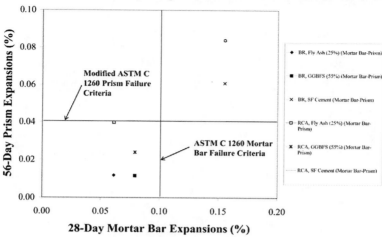

Figure 6: Modified ASTM C 1260 56-Day Prism versus ASTM C 1260 28-Day Mortar Bar Results

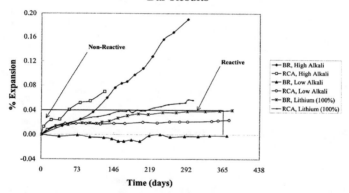

Figure 7: ASTM C 1293 Prism (Low Alkali Cement and Lithium Nitrate

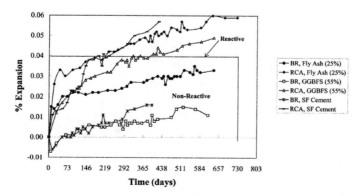

Figure 8: ASTM C 1293 Prism (Cement Substitution Mitigations)

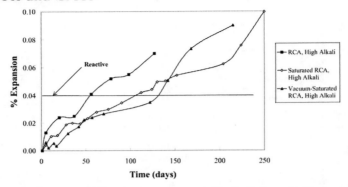

Figure 9: ASTM C 1293 Prism (RCA/High Alkali Cement Saturation)

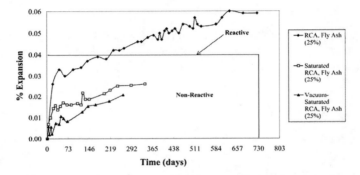

Figure 10: ASTM C 1293 Prism (RCA/Class F Fly Ash Saturation)

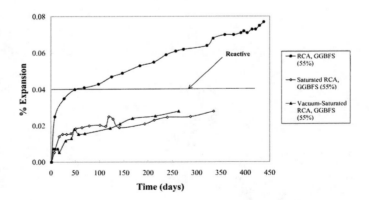

Figure 11: ASTM C 1293 Prism (RCA/GGBFS Saturation)

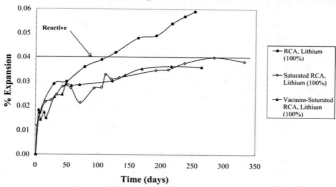

Figure 12: ASTM C 1293 Prism (RCA/Lithium Nitrate Saturation)

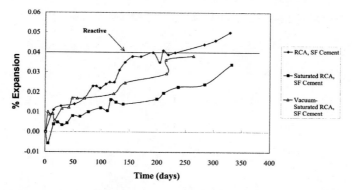

Figure 13: ASTM C 1293 Prism (RCA/Silica Fume Blended Cement Saturation)

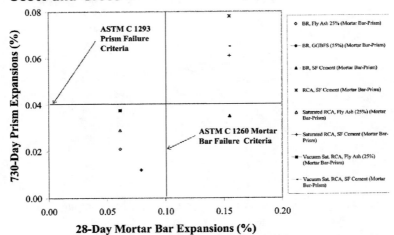

Figure 14: ASTM C 1293 730-Day Prism versus ASTM C 1260 28-Day Mortar Bar Results

Use of Recycled Glass as Aggregate for Architectural Concrete

by C. Meyer and S. Shimanovich

Synopsis: Secondary markets for waste glass have been widely developed in Europe. In the United States, on the other hand, most post-consumer glass is still being land-filled, primarily because it is mixed-color. Also, the need to clean the often highly contaminated glass constitutes a barrier against its beneficial use. For several years, an ongoing research effort at Columbia University has explored the potential of waste glass as an aggregate for concrete. The primary technical problem, caused by alkali-silica reaction, can be solved with existing means. The economics of beneficiating recycled glass in large urban areas such as New York City constitutes a more formidable barrier, because the standard aggregate that the glass would replace, whether sand or gravel, is relatively inexpensive. However, if the esthetic potential of color-sorted glass is exploited fully, the economic picture changes, and glass processors are more likely to create the link between curbside collection and concrete producer. A number of architectural concrete applications have been explored to date, and design professionals and developers have shown keen interest in adding high-quality concrete products to their palette of options. Some of these are already produced commercially. By developing a promising secondary market for recycled glass as a value-added component of architectural concrete, it is possible to offer the concrete industry new directions to shed its image of being environmentally unfriendly and to actively embrace the principles of sustainable development.

Keywords: aggregate; alkali-silica reaction (ASR); architectural concrete; glass; recycling; sustainable development

78   Meyer and Shimanovich

Christian Meyer is a Professor of Civil Engineering at Columbia University in the City of New York. His interests are in structural engineering, structural concrete, and concrete material science and technology. He is member of ACI Committees 446, 447, 554, and currently chairs Committee 555, Concrete with Recycled Materials.

Semyon Shimanovich is a Senior Research Scientist in the Department of Civil Engineering and Engineering Mechanics at Columbia University. His interests are in concrete materials science and technology, building materials for floors, coatings, precast high strength and self-leveling concrete, and concrete with recycled and waste materials.

INTRODUCTION

Compared with many other developed countries, recycling and the reuse of natural resources have traditionally been of low priority and often nonexistent in the U.S. This is no longer the case. A dramatic change in attitude can be felt today throughout the country. This change came relatively suddenly, gaining significant momentum in the early 1970s. A key event was the celebration of Earth Day in 1970, when a large part of the American public became aware of the limits of the nation's resources and grew concerned about the deteriorating environment, whether soil, water, or air and realized that our planet is indeed finite in size and in resources.

One other reason why the American public was so slow in realizing the finiteness of its resources was the size of the country. Unlike in many countries in Europe, for example, it seemed as if there was plenty of space to dump its refuse and waste material. This perception has definitely changed, when reality needed to be faced. Not only did the physical space available for landfills become sparse. Also legislation on the federal, state, and local levels imposed severe environmental restrictions on them. As a result, many existing landfills had to be closed, with costly cleanup measures needed for some, and it is now becoming increasingly difficult to open up new landfills.

These developments are nowhere as dramatic as in New York City, which probably generates more solid waste than any other city in the world, including those with much larger populations. The serious problem of solid waste disposal became even more critical, when the Freshkill Landfill on Staten Island, the world's largest, had to be closed. It does not reflect positively on the City's administration that the main alternate option pursued so far is to ship the garbage out of state to the lowest bidder.

Until recently, the City had a recycling program, under which metals, plastics, and glass were collected intermingled and turned over to Materials Recycling Facilities (MRF's) for a contracted price. The MRFs were able to sell metals and much of the plastics. But the glass was usually land-filled, because there was no secondary market for the cullet, which was of mixed color and generally highly contaminated. When the costs for the recycled material escalated beyond $100 per ton, the new City administration simply stopped its entire recycling program. It is now in the process of entering new contracts to bring the costs under control. Glass constitutes approximately 6% of New York City's solid waste, and at present, its disposal is estimated to cost taxpayers as much as $60 million each year. Much of this money could be saved by beneficiating the glass as aggregate for concrete products, not counting the environmental benefits associated with such recycling.

GLASS AS CONCRETE AGGREGATE

Conventional wisdom has it that glass is not suitable for use in concrete because the alkali in the cement paste may react with the silica in the glass and, in the presence of moisture, can cause damaging swelling. This alkali-silica reaction (ASR) is not limited to glass aggregate but can occur in concrete produced with many other types of aggregate as well and therefore poses a serious problem for the concrete industry [1]. Whereas ASR-induced damage caused by most natural reactive aggregates is a long-term phenomenon, which may take years to manifest itself, and is subject to considerable uncertainty, the reactions involving soda lime glass that is commonly used for beverage containers have the "advantage" of being quite certain and comparably rapid. For this reason, such glass lends itself to laboratory studies of ASR and as the material of choice for developing methods to avoid or minimize ASR-induced damage.

Rather than yielding to conventional wisdom by avoiding the use of glass in concrete, researchers might choose to regard the ASR-problem as a challenge to be overcome. This is the reason why in 1995, a major research project was initiated at Columbia University with support from the New York State Energy Research and Development Authority (NYSERDA) to investigate the feasibility of using waste glass in concrete products [2-4].

There are several ways to avoid ASR or its damaging effects.

- Grind the glass to pass at least mesh size #50 [4].
- Mineral admixtures (such as metakaolin, fly ash, slag, etc), added in appropriate amounts are known to effectively reduce expansions [5-8].
- The glass can be made alkali-resistant, for example, by coating it with zirconium – a solution chosen by the glass fiber industry, but impractical for post-consumer waste glass.
- Similarly, the glass chemistry can be modified, for example by adding elements to the glass melt which are known to reduce or eliminate ASR-induced expansions.
- Because ASR needs three factors to thrive (alkali, silica, and moisture), sealing the concrete to keep it dry can minimize or avoid the problem.
- Low-alkali cements can be effective, as long as alkalis from the environment are kept away.
- Special ASR-resistant cements may be developed and are already being offered commercially.

It should be cautioned, however, that ASR is an extremely complex phenomenon. Even small changes in the glass chemistry can have profound consequences. There are considerable differences between consumer bottle glass, window glass, windshields, and light bulbs, so that each glass source needs to be evaluated carefully in order to assure the quality and durability of the concrete end product.

Concrete with glass aggregate is different from regular concrete in several respects. Mix designs for concrete products that are mass-produced in automated facilities need to be optimized for the particular production equipment used. For example, the zero water absorption of glass improves the mix rheology that influences the mix design, whether a dry or wet process is used. The key to success lies largely in the special admixtures that optimize the flow and consolidation properties without impairing the mechanical and durability properties of the hardened concrete, while being chemically compatible with the cement.

Aside from the ASR problem, glass offers a number of advantages, if used as aggregate in concrete. These can be summarized as follows:

- Primarily because of its basically zero water absorption, glass is one of the most durable materials known to man. With the current emphasis on durability of high-performance concrete, it is only natural to rely on extremely durable ingredients.
- The excellent hardness of glass gives the concrete an abrasion resistance that very few natural stone aggregates can match.
- For a number of reasons, glass aggregate improves the flow properties of fresh concrete so that very high strengths can be obtained even without the use of superplasticizers.
- The esthetic potential of color-sorted post-consumer glass, not to mention specialty glass, has barely been explored at all and offers novel opportunities for design professionals.
- Very finely ground glass has pozzolanic properties and therefore can serve both as partial cement replacement and filler.

PRODUCT DEVELOPMENT

When discussing the economics of glass as concrete aggregate, it is useful to draw a distinction between *commodity products* and *value-added products*. The main purpose of using crushed glass in commodity products is to divert as much glass as possible from the waste stream into beneficial use applications. However, the markets for commodity products, such as paving stones and concrete masonry units are typically very competitive, with low profit margins. Therefore, the economic benefit of substituting glass for fine aggregate or even cement is marginal at best, because it is essential that a dependable source of glass be available that is clean, crushed and graded to specification. Moreover, if the glass is ground sufficiently fine to minimize the potential of ASR damage, it is not possible to see whether products contain such glass or not.

The first commodity product to be developed for commercial production was a concrete masonry block unit, for which a rather modest 10% of fine aggregate was replaced by finely ground glass, or 10% of the cement by glass powder. In view of such small amounts of material substitution, no major effects on strength or other block properties were observed, as expected.

In value-added products, the purpose of the glass substitution is to exploit the special properties of the glass and thereby add value to a material that otherwise would be a waste product. If the glass is sorted by color and this is coordinated with the color of the cement matrix, novel esthetic effects can be achieved, which can be further enhanced with appropriate surface treatments. Surface textures can range from highly polished surfaces, for example, for tiles or tabletop counters, to exposed aggregate surfaces for building façade elements. The economic value of the glass derives to some extent from the material it replaces. For example, terrazzo tiles often contain costly natural aggregate such as imported marble chips. But it is possible to produce with glass visual effects that cannot be achieved with any other material. This would leave the end products basically without competition and, therefore, very attractive for producers and users alike. Glass concrete terrazzo tiles are already being manufactured commercially by Wausau Tile, Inc., of Wausau, WI. Photographs of sample tiles are not shown here, because black-and-white reproductions cannot do justice to their actual appearance.

It should also be mentioned that plain glass concrete is just as brittle as regular concrete. For this reason it may be advantageous to reinforce glass concrete products with either randomly distributed short fibers or, in the case of thin sheets or panels, with fibermesh or textile reinforcement.

One product, which is close to being mass-produced, is a paving stone with up to 100% glass aggregate. Its appeal lies in the novel colors and surface texture effects, such as special light reflections, that cannot be obtained with regular natural aggregate. Other advantages are the greatly reduced water absorption and excellent abrasion resistance due to the high hardness of glass. The paving stone can also be reinforced with fibers to improve its mechanical properties, especially its energy absorption capacity and fracture toughness. A fiber-reinforced paver is just as likely as any other paver to crack under impact, but the fibers will keep such cracks so small as to be basically invisible. Initial tests have shown that the freeze-thaw cycle resistance is excellent, with barely any damage after 600 cycles.

Architects appreciate the many novel surface textures and color effects offered by glass aggregate. This is particularly true for exposed aggregate technologies, which have been known in the architectural concrete community for some time. The added value derives from the fact that both regular concrete and waste glass are inexpensive, but if used in combination, these two component materials can fetch a price that is only marginally controlled by the costs of production. Since most alternative materials are more costly, glass concrete façade elements offer architects and other design professionals considerable flexibility.

The special effects that can be achieved with glass aggregate in the various architectural and decorative fields can be stunning, and the number of potential applications is limited only by one's imagination. To name just a few:

- Building façade elements
- Precast wall panels

- Partitions
- Floor tiles
- Wall tiles and panels
- Elevator paneling
- Table top counters
- Park benches
- Planters
- Trash receptacles

ECONOMICS OF GLASS RECYCLING

The economics of recycled materials is to some extent different from that of virgin materials. Principally, the laws of supply and demand apply to both. On the one side of the economic equation are real costs associated with supplying a commodity, and on the other side is the price that the demand for it generates on the open market. But there are other factors, which complicate an economic analysis of recycled glass.

The largest cost factor is associated with collection, which varies widely among different municipalities. In New York City there are unique reasons why the cost of separate curbside collection of recyclables is unusually high. After a previous recycling program had been discontinued because of escalating costs, the Bloomberg Administration is at present (early 2003) in the process of designing a new recycling program. In the interim period, the glass is simply added to the common solid waste stream, which places a burden on taxpayers, which is unjustifiable, especially during a period of substantial budgetary deficits. The City generates probably more solid waste than any other and at the same time has no nearby landfills to dispose of it. This unique situation is the primary reason for the unusually large negative value of waste glass at the source.

Compared with the cost of curbside collection, the various processing costs are moderate. The glass needs to be cleaned to remove organics, sugars, and other deleterious substances such as bottle caps and labels. Crushing and grading of the cullet to specification is relatively inexpensive and should not cost more than about $1 or $2 per ton. But the cost of transportation can become prohibitive when large distances are involved. The main challenge for a concrete producer is to contractually lock in a secure supply of glass for a guaranteed price.

The value of the glass to the producer or more precisely the amount he is willing to pay for it depends on the material to be replaced. Whereas regular sand and gravel are very inexpensive (typically of the order of $10 to 15 per ton), a producer may be willing to pay hundreds of dollars per ton for specialty aggregates, if the value added to the end product justifies it, i.e., can be passed on to the consumer. By identifying the special properties of crushed glass and exploiting these in the design of concrete products, it is possible to add value to the material, if the result is a product with properties that are superior to those produced with natural aggregates. This added value needs to be emphasized when marketing the product. Moreover, the special esthetic qualities of glass can produce effects that cannot be achieved with natural aggregate. This makes the glass basically without equal for selected applications, so that its value is determined almost entirely by how much the end user is willing to pay.

In addition, there is the possibility of government intervention. If the public at large fails to recognize the need to conserve resources, its elected representatives are expected to act in the public's best interest. In fact, many state and local authorities are already actively intervening in the market economy with incentives and disincentives or prescriptive legislation. Incentives may be offered in the form of tax breaks for developers, who utilize a certain percentage of recycled material content, and special legislation may require certain projects to adhere to principles of "green building" design, for example, certification under the U.S. Green Building Council's LEED program [5]. In exceptional cases, an environmentally concerned developer may decide to pay a premium for environmentally friendly materials, but concrete producers cannot count on this. It is preferable to rely on the superior properties inherent in the glass and the added value of a better product and then let the forces of supply and demand determine the economics.

The creation of a new secondary market for glass cullet can increase the demand for the glass and according to the law of supply and demand lead to an increase in price. From an environmental policy viewpoint such a development would be desirable, and municipalities would end up paying less for the disposal of the glass. The effect on concrete producers would be that only the higher-end products are likely to utilize glass aggregate because of its increased value.

CONCLUSION

Concrete is the world's most important building material, and it is now possible to engineer its properties to satisfy almost any reasonable performance specifications, whether these pertain to mechanical, thermal, chemical, or durability requirements, or relate to unique esthetic demands. By using waste glass as aggregate in concrete, several benefits can be obtained simultaneously. Because of its inherent chemical and physical characteristics, glass can improve the mechanical and other properties of concrete products. Moreover, its esthetic qualities can be exploited to produce decorative effects that are not possible to create with natural aggregates. The glass needs to be cleaned, crushed and graded to specification. Color-sorting adds additional value.

Some glass products such as terrazzo tiles and paving stones are already being mass-produced commercially. The economic success depends on how effectively these products can be marketed. The changing public attitudes towards environmental issues as well as governmental intervention through incentives and disincentives as well as outright prescriptive legislation are indications that such products have a high probability of success. Moreover, the development of concrete with glass aggregate can serve as an example for the industry to reduce its impact on the environment and embrace the principles of sustainable development.

Academic research can lead to new technologies. However, for such technologies to become commercially successful, proper technology transfer is needed as well as adequate funding, that makes the research effort possible in the first place. This requires cooperation between industry and academia. The success story of glass concrete is an illustration that investments in research do not have to be large for the results to be lucrative.

REFERENCES

1. Fournier, B. et al, eds., "Alkali-aggregate reaction in concrete", Proc., 11th Int. Conf. On Alkali-Aggregate Reaction, Quebec City, QC, CRIB, 2000.
2. Meyer, C. and S. Baxter, "Use of Recycled Glass for Concrete Masonry Blocks", Final Report to New York State Energy Research and Development Authority, Rep. No. 97-15, Albany, NY, Nov. 1997.
3. Jin, W., "Alkali-Silica Reaction in Concrete with Glass Aggregate – A Chemo-Physico-Mechanical Approach", Ph.D. Dissertation, Columbia University, New York, NY, 1998.
4. Jin, W., C. Meyer and S. Baxter, "Glascrete – Concrete with Glass Aggregate", ACI Materials Journal, March-April 2000.
5. Sabir, Wild, and Bai, "Metakaolin and Calcined Clays as Pozzolans for Concrete. A Review." Cement & Concrete Composites 23:441-454, 2001.
6. ACI Committee 232, "Use of Fly Ash in Concrete", Technical Report 232.2R-96, American Concrete Institute, 1996.
7. ACI Committee 233, "Ground Granulated Blast-Furnace Slag as a Cementitious Constituent in Concrete", Technical Report 233R-95, American Concrete Institute, 1995.
8. ACI Committee 234, "Guide for the Use of Silica Fume in Concrete", Technical Report 234R-96, American Concrete Institute, 1996.
9. U.S. Green Building Council, LEED Certification Program, Washington, D.C.

SP-219—7

Properties of Flowable Slurry Containing Wood Ash

by T. R. Naik, R. N. Kraus, Y. Chun, and R. Siddique

Synopsis: Three series of flowable slurry mixtures were made, each series with three different sources of wood ash (W-1, W-2, and W-3). The series of mixtures were: low-strength (0.3 to 0.7 MPa), medium-strength (0.7 to 3.5 MPa), and high-strength (3.5 to 8 MPa) mixtures. Tests were performed for flow, air content, unit weight, bleeding, settlement, compressive strength, and water permeability. Wood ashes W-1 and W-3 caused expansive reactions in CLSM mixtures resulting in little or slight (average 1%) net shrinkage of CLSM. Wood ash W-2 caused either significant net swelling (15% for Mixture 2-L, and 21% for Mixture 2-M) or no shrinkage (Mixture 2-H) of CLSM. The 91-day compressive strength of low-strength, medium-strength, and high-strength slurry mixtures was in the ranges of 0.38 to 0.97 MPa, 1.59 to 5.28 MPa, and 4.00 to 8.62 MPa, respectively. Overall, the slurry mixtures showed an average increase in strength of 150% (range: 25% to 450%) between the ages of 28 days and 91 days. This was attributed to pozzolanic and cementitious reactions of wood ash. In general, water permeability of CLSM mixtures decreased with age.

Keywords: bleed water; cement; compressive strength; controlled low-strength materials (CLSM); flowable slurry; permeability; wood ash

ACI Member, Tarun R. Naik, FACI, is Professor of Structural Engineering and Director of the UWM Center for By-Products Utilization (UWM-CBU), Department of Civil Engineering and Mechanics at the University of Wisconsin - Milwaukee. He is a member of ACI Committees 555, "Concrete with Recycled Materials"; 232, "Fly Ash and Natural Pozzolans in Concrete"; 229, "CLSM"; 214, "Evaluation of Results of Tests Used to Determine the Strength of Concrete"; and 123, "Research". He was also chairman of the ASCE Technical Committee on "Emerging Materials" (1995-2000).

Rudolph N. Kraus, MACI, is Assistant Director of the UWM-CBU. He has been involved with numerous projects on the use of by-product materials including utilization of used foundry sand and wood and coal fly ash in controlled low-strength materials (CLSM), evaluation and development of CLSM, evaluation of lightweight aggregates, and use of by-product materials in the production of cast-concrete products.

Yoon-moon Chun is a Research Associate at the UWM-CBU. His research interests include use of industrial by-products such as coal ash and used foundry sand in construction materials, especially use of fibrous residuals from pulp and paper mills in concrete.

Rafat Siddique is a Research Associate at the UWM-CBU. His research interests are high-volume fly ash concrete, high-strength high-performance concrete, and use of by-product materials in construction. He is on leave from the Thapar Institute of Engineering & Technology, Patiala, India, where he is a Senior Assistant Professor.

INTRODUCTION

U.S. pulp and paper mills generate about one million dry tonne of wood ash per year. National Council for Air and Stream Improvement (NCASI) has estimated that of the total wood ash, only about one-third is being utilized (1). The disposal of this large-scale generation of wood ash is a major problem for the industry, mainly pulp mills, saw mills, and energy generating plants that utilize wood and wood residue. The problem concerning the disposal of wood ash in landfills is accentuated by potentially limited landfill space available, strict environmental regulations, and high costs. Co-firing wood residue with coal or other fuels lead to regulatory differentiation between ash from wood residue alone and ash from wood mixed with coal and/or other fuels. Therefore, beneficial utilization options for wood ash is essential for the industry. One of the possible uses of wood ash is in the production of controlled low-strength materials (CLSM), also widely known as flowable slurry. CLSM is a high-fluidity cementitious material that flows like a liquid, self-levels without compacting, and supports as a solid when hardened. The American Concrete Institute (2) describes CLSM as a cementitious material that is in a flowable state at the time of placement and has a specified compressive strength of 8.3 MPa (1200 psi) or less. A number of names including flowable fill, unshrinkable fill, manufactured soil, controlled-density fill, and flowable mortar are being used to describe this material. CLSM is used primarily for non-structural and light-structural applications such as backfills, sound insulating and isolation fills, pavement bases, conduit bedding, erosion control, and void filling (2).

Higher-strength CLSM can be used in applications where future excavation is unlikely, such as structural fill under buildings (2). In deciding mixture proportions of CLSM, factors such as flowability, strength, and excavatability are evaluated. Permeability is, for many uses, an important property of CLSM. Permeability of CLSM depends on mixture proportions, properties of constituent materials, water-cementitious material ratio (w/cm), and age.

RESEARCH SIGNIFICANCE

The principal objective of this investigation was to evaluate the strength and permeability of flowable slurry incorporating wood ash. In this investigation, three series of slurry mixtures—low-strength (0.3 to 0.7 MPa), medium-strength (0.7 to 3.5 MPa), and high-strength (3.5 to 8 MPa)—were proportioned, each series from three different sources of wood ash (W-1, W-2, and W-3). The results of this investigation will establish mixture proportions and production technology for flowable slurry containing wood ash.

LITERATURE REVIEW

Numerous studies by Naik and his associates (3-9), Ramme et al. (10), Krell (11), Swaffer and Price (12), Larson (13,14), and Fuston et al. (15) have examined and reported on flowable slurry properties such as density, strength, settlement, permeability, shrinkage, and other properties. Lai (16) reported that the compressive strength of flowable mortars containing high-volume coal ash is applicable for backfill or base course construction. It was mentioned that the 28-day compressive strength of about 1 MPa could be achieved with 6% (by mass) cement at excellent flowability. In the 1980s, Naik et al. (9) developed excavatable CLSM mixtures having compressive strength between 0.3 and 0.7 MPa (50 and 100 psi) at 28 days. Naik and Singh (5) have reported on strength and water permeability of slurry materials (0.3 to 0.7 MPa) containing used foundry sand and fly ash as 3×10^{-6} to 74×10^{-6} cm/s. Tikalsky et al. also (17) evaluated the potential of used foundry sand as a constituent of controlled low-strength material (CLSM), and concluded that used foundry sand provides a high-quality material for CLSM. Tikalsky et al. (18) evaluated CLSM containing clay-bonded and chemically-bonded used foundry sand, and showed that used foundry sand can be successfully used in CLSM. Horiguchi et al. (19) evaluated the potential use of high-carbon (12% loss on ignition [LOI]) fly ash plus non-standard bottom ash in CLSM. A total of 20 mixtures were tested for flowability, bleeding, and short-term and long-term compressive strengths. Test results indicated that there is an optimum fly ash to bottom ash ratio for the desirable physical properties of CLSM. CLSM with high-carbon fly ash and non-standard bottom ash showed excellent performance indicating ecological and economical applicability to CLSM. Horiguchi et al. (20) also investigated physical and durability characteristics of CLSM made with used foundry sand and bottom ash as fine aggregates. Based on the test results, they concluded that the one-dimensional frost-heaving rate of CLSM made with used foundry sand and bottom ash ranged from 0.4% to 1.8%. Naik et al. (21) developed two types of CLSM utilizing post-consumer glass aggregate and fly ash. One group of CLSM consisted of cement, fly ash, glass, and water; and, another group of CLSM consisted of cement, sand, glass, and water. They concluded that the flowable slurries containing glass satisfied ACI Committee 229 recommendations (2). Gassman et al. (22) examined the effects of prolonged mixing and re-tempering on the fluid- and hardened-state properties of CLSM. The test results showed that extending the mixing time beyond 30 minutes decreased unconfined compressive strength and delayed the time of setting. Re-tempering did not affect the 28-day strength; however, it did affect the 90-day strength depending upon the mixing time. There are insufficient data available for the permeability of slurry materials. Furthermore, there is a general lack of any information available for use of wood ash in flowable slurry.

EXPERIMENTAL PROGRAM

Materials

Type I portland cement conforming to ASTM C 150 requirements was used in this investigation. The fine aggregate used was natural sand having dry-rodded unit weight of 1770 kg/m^3, specific gravity of 2.67, absorption of 1.3%, and fineness modulus of 2.7. Three different sources of wood ash were used, and these sources are designated as W1, W2, and W3. W1 was obtained from Niagara, Wisconsin; W2 from Biron, Wisconsin; and W3 from Rothschild, Wisconsin. Physical properties and chemical composition of the three sources of wood ash are presented in Tables 1 and 2, respectively. The wood ash used in this project did not meet all the requirements of ASTM C 618 for coal ash (Class C and F) or natural pozzolan (Class N).

Each source of wood ash exhibited different physical properties (Table 1). Fineness of the wood ash (% retained on 45 μm sieve) varied from 23% to 90%. Source W-1 met the ASTM requirement for fineness (34% maximum), while sources W-2 and W-3 exceeded the ASTM limit. Small charcoal pieces were also visible in Source W-2. Source W-3 was very coarse with a particle size distribution similar to coarse sand. The strength activity index of the wood ash is a comparison of the compressive strength development of 50 mm mortar cubes that have 20% (by mass) replacement of cement with wood ash, with compressive strength of standard cement mortar. Wood ashes W-1 and W-3 met the strength activity index requirement of ASTM (75% minimum at either 7 or 28 days), while wood ash W-2 did not meet the requirement. Water requirement of all wood ashes also exceeded the maximum water requirement for coal ash specified by ASTM C 618 (105%); however, Sources W-1 and W-3 satisfied the requirement for natural pozzolan (Class N). The higher water requirement indicated that for concrete and CLSM containing wood ash, more water would be required to produce the same slump or flow as compared with the control mixture. Unit weight values of the wood ashes W-1 and W-2 were 550 and 410 kg/m^3, respectively. These unit weights are significantly less than the unit weight of a typical ASTM Class C or Class F fly ash (approximately 1100 to 1300 kg/m^3). Source W-3 had a unit weight of 1380 kg/m^3. Specific gravity of wood ash Sources W-1, W-2, and W-3 range from 2.26 to 2.60. Specific gravity of wood ash Source W-1 was lower than that of a typical coal fly ash (approximately 2.40 to 2.60). The low unit weight and specific gravity of some wood ashes indicate that adding these wood ashes into CLSM would lower the unit weight of CLSM. This may be useful for obtaining lighter-weight CLSM.

The results of the chemical analysis of the wood ashes are given in Table 2. All wood ashes did not meet all the chemical requirements of ASTM C 618, particularly for the amount of carbon as shown by LOI test results. The LOI obtained for the wood ashes range from 6.7% to 58.1%. These high LOI ashes probably will present some difficulties when developing air-entrained concrete mixtures. The higher carbon content tends to reduce the amount of air entrained in the concrete mixture and thus requires higher dosages of air-entraining admixtures. However, the higher carbon contents of the wood ashes should not affect the performance of these ashes in CLSM. Wood ash W-2 showed a very low value of $SiO_2 + Al_2O_3 + Fe_2O_3$ (23.4%) most likely due to its high LOI. Wood ashes W-2 and W-3 showed high lime contents (13.7% and 19.6%, respectively).

Mixture Proportions

Three series of slurry mixtures were proportioned for the three different sources of wood ash. Each of the three series was developed to obtain a different long-term compressive strength level using the three sources of wood ash. The three target long-term strength levels developed for the project were 0.3 to 0.7 MPa (low-strength), 0.7 to 3.5 MPa (medium-strength), and 3.5 to 8 MPa (high-strength). The low-strength CLSM mixtures (1-L, 2-L, and 3-L) consisted of cement, wood ash, and water. The medium-strength CLSM mixtures (1-M, 2-M, and 3-M) consisted of an increased amount of cement, wood ash, and water. The high-strength CLSM mixtures (1-H, 2-H, and 3-H) consisted of cement, wood ash, sand, and water. These CLSM mixtures were proportioned to maintain a practical value of flow in the range of approximately 250 ± 50 mm. Mixture proportions for low-strength, medium-strength, and high-strength slurry mixtures are presented in Table 3. Cement content of the slurry mixtures was: 53 to 89 kg/m^3 for the low-strength mixtures; 101 to 228 kg/m^3 for the medium-strength mixtures; and, 169 to 205 kg/m^3 for the high-strength mixtures. Within each series, the mixture containing wood ash W-2 was proportioned to have the highest amount of cement and lowest amount of wood ash because of the low strength activity index, high water requirement, and high LOI of wood ash W-2 (Tables 1 and 2).

Manufacturing Technique

The flowable slurry was mixed in a 0.25-m^3 rotating drum mixer. The slurry ingredients were weighed and loaded in the mixer in the following manner. Initially, the required amount of cement and half the amount of wood ash and sand (for high-strength mixtures) was loaded in the mixer and mixed for three minutes. About three quarters of the specified water was then added to the mixer and mixed for an additional three minutes. The remaining wood ash, sand (when used), and water were added to the mixer and mixed for an additional five minutes. Additional water was added in smaller quantities to ensure reaching the required flow consistency of CLSM.

Preparation and Testing of Specimens

The flow/spread, air content, temperature, and unit weight were determined for each mixture before casting test specimens. Fresh slurry was tested for flow/spread in accordance with ASTM Standard Test Method for Flow Consistency of Controlled Low Strength Material (CLSM) (D 6103). The temperature of the slurry and ambient air temperature were measured and recorded. The unit weight and air content were determined in accordance with ASTM Standard Test Method for Unit Weight, Yield, Cement Content, and Air Content (Gravimetric) of Controlled Low Strength Material (CLSM) (D 6023). Measurement of bleed water and settlement of the flowable slurry was done using slurry cast in 150 × 300 mm cylindrical plastic molds. The compressive strength test specimens were cast in 150 × 300 mm cylindrical plastic molds. The permeability specimens were cast in 100 × 125 mm cylindrical plastic molds. All specimens were prepared in accordance with ASTM C 192. The strength specimens and permeability specimens were typically cured for one day in their molds in the laboratory at about 23 ± 3°C. They were then minimum placed in a standard moist-curing room maintained at 100% relative humidity and 23 ± 3°C temperature until the time of test. Compressive strength was determined in accordance with ASTM Standard Test Method for Preparation and Testing of Controlled Low Strength Material (CLSM) Test Cylinders (D 4832) by testing three cylinders per CLSM mixture at 3, 7, 28, and 91 days. The permeability of the slurry was determined at 28 and 91 days in accordance with ASTM Standard Test Methods for Measurement of Hydraulic Conductivity of Saturated Porous Materials Using a Flexible Wall Permeameter (D 5084) by testing two or three specimens per CLSM mixture, per test age.

TEST RESULTS AND ANALYSIS

Flow/Spread, W/Cm, Air Content, Temperature, and Unit Weight

Overall, flow/spread of the slurry mixtures was in the range of about 240 to 270 mm (Table 3). Within each series of slurry mixtures, the mixture containing wood ash W-3 showed the lowest w/cm (in this case, $w/[c+wa]$), followed by the mixture containing wood ash W-1 and the mixture containing wood ash W-2 for achieving desired flow/spread. Entrapped air content was in the range of 1.4% to 6.3%. Within each series of slurry mixtures, the mixture containing wood ash W-1 showed the lowest air content, followed by the mixture containing wood ash W-3 and the mixture containing wood ash W-2. The unit weight of the slurry material was found to be in the range of 1190 and 2030 kg/m^3. Within each Series of slurry mixtures, the mixture with wood ash W-3 showed the highest unit weight, followed by the mixture with wood ash W-1 and the mixture W-2. This is in agreement with the w/cm of the slurry mixtures, higher water demand, and LOI of the different wood ashes.

Approximately the same quantity of water was required for the medium-strength CLSM mixtures as used for the low-strength CLSM mixtures. This was expected since only the quantity of cement was increased for these mixtures. The high-strength CLSM mixtures contained fine aggregate (concrete sand) to develop the higher CLSM strength. Quantities of water required to obtain the design flow of approximately 250 mm decreased from those of the mixtures that did not contain sand. Approximately 360 to 430 kg of water per m^3 of slurry was required to reach the target flow for the high-strength mixtures, compared with 480 to 640 kg of water per m^3 of slurry for the low- or medium-strength mixtures made without sand. Unit weight of the high-strength CLSM mixtures was also significantly higher than the low-strength mixtures and medium-strength mixtures. This was attributed to the use of sand in high-strength slurry mixtures. Sand had significantly higher unit weight and specific gravity than the wood ashes.

Bleed Water

Bleeding was measured as the depth of water above the CLSM mixture level in a 150 × 300 mm plastic cylindrical mold. The bleed water gives an indication of the cohesiveness of the CLSM mixture. The initial bleed water at one-hour for the Mixtures 1-L, 2-L, and 3-L was 10, 3, and 10 mm, respectively. For Mixtures 1-L and 2-L, bleed water evaporated in 8 hours, whereas Mixture 3-L recorded up to 1.5 mm of bleed water until 3 days. The one-hour bleed water depth for the Mixtures 1-M, 2-M, and 3-M was 13, 0, and 5 mm, respectively. Evaporation of bleed water took place in 3 days for Mixture 1-M, and 8 hours for Mixture 3-M. The one-hour bleed water depth for the Mixtures 1-H, 2-H, and 3-H was 5, 10, and 10 mm, respectively. Bleed water evaporated in 24 hours from the top of Mixtures 1-H, 2-H, and 3-H. Bleed water is closely related to settlement of slurry.

Settlement

The settlement of the CLSM was determined by measuring the level of the CLSM in a 150 × 300 mm cylinder with reference to the top of the cylinder. The settlement measurements would indicate any potential of the CLSM to shrink or expand, and possible cracks that may develop on the surface upon drying which may lead to the inflow of water into the CLSM. The results are presented in Table 4. Overall, all of the CLSM mixtures (except Mixtures 2-L and 2-M) showed shrinkage initially at one hour. But, wood ashes W-1 and W-3 caused some degree of swelling of slurry mixtures, and the net or final settlement of the CLSM mixtures containing these wood ashes was relatively very low (average 3 mm [range 0 to 10 mm]), which is very desirable. The CLSM mixtures containing wood ash W-2 showed either significant net swelling (46 mm [15%] for Mixture 2-L, and 62 mm [21%] for Mixture 2-M) or no net shrinkage (Mixture 2-H). Mixtures 2-L and 2-M showed expansion from the beginning, while Mixture 2-H exhibited shrinkage initially. Mixture 2-H contained about half as much wood ash as either Mixture 2-L or 2-M (Table 3). Also, the presence of sand in Mixture 2-H and high unit weight of Mixture 2-H might have provided a confining effect. Precautions should be taken when using Mixtures 2-L and 2-M due to the significant expansion. Placing these mixtures in a confined volume could lead to internal pressure that would have to be accounted for in the design. However, the expansive characteristics may be useful when filling a space such as an abandoned tank, pipe, or tunnel to assure that no voids remain inside.

Compressive Strength

The low-strength slurry mixtures were proportioned to have a long-term compressive strength between 0.3 and 0.7 MPa. Mixtures 1-L and 3-L achieved the desired strength level before reaching 28 days, and Mixture 2-L before reaching 91 days

(Fig. 1). The respective compressive strength of Mixtures 1-L, 2-L, and 3-L was 0.38, 0.24, and 0.41 MPa at 28 days, and increased to 0.48, 0.38, and 0.97 MPa at 91 days. This is an increase of 27%, 57%, and 133%, respectively (average 70%), in the compressive strength between the ages of 28 days and 91 days. At 91 days, Mixture 3-L containing wood ash W-3 exceeded the desirable long-term maximum strength of 0.7 MPa, and would have shown a significant increase in strength if test continued beyond 91 days (Fig. 1). This indicates that the amount of cement used for Mixture 3-L can be further reduced in the future. Among the low-strength slurry mixtures, Mixture 2-L showed the lowest strength up to the test age of 91 days. This was attributed to the high w/cm and low unit weight of slurry Mixture 2-L. However, Mixture 2-L still achieved the desired level of long-term strength and probably would have shown a significant long-term strength gain if test continued.

The medium-strength slurry mixtures were proportioned to have a long-term compressive strength between 0.7 and 3.5 MPa. All the medium-strength slurry mixtures achieved the desired strength level before reaching 28 days (Fig. 2). The respective compressive strength of Mixtures 1-M, 2-M, and 3-M was 1.38, 0.97, and 1.28 MPa at 28 days, and increased to 2.69, 5.28, and 1.59 MPa at 91 days. This is an increase of 95%, 446%, and 24 %, respectively (average 190%), in strength at 91 days compared with 28 days.

All the high-strength slurry mixtures achieved desired long-term strength level (3.5 and 8 MPa) before reaching 91 days (Fig. 3). The respective compressive strength of Mixtures 1-H, 2-H, and 3-H was 3.24, 1.59, and 3.10 MPa at 28 days, and increased to 6.66, 8.62, and 4.00 MPa at 91 days. This is an increase of 105%, 443%, and 29%, respectively (average 190%), in strength between the ages of 28 days and 91 days. The low strength of Mixture 2-H at 28 days was attributed to its high w/cm, high air content, and low unit weight (Table 3).

In summary, all the slurry mixtures achieved the desired long-term strength levels. All of the slurry mixtures made with wood ash exhibited a significant increase (average 150% [range 24% to 446%]) in compressive strength between the ages of 28 days and 91 days. This indicates that continuing pozzolonic as well as cementitious reactions occur in the CLSM mixtures when wood ash is used even with high w/cm. Mixtures 3-L, 1-M, 2-M, 1-H, and 2-H either exceeded the desired long-term strength or would have shown as such if test continued beyond 91 days. The quantities of cement in these mixtures could be reduced. The significant long-term gain in compressive strength of the CLSM containing wood ash should be accounted for when designing for the needs of various applications. In future investigations, it would be highly desirable to determine the compressive strength of the CLSM containing wood ash, beyond 91 days (for example, 6 months and 1 year).

Permeability

The water permeability test results for the CLSM mixtures are given in Table 5. Overall, correlation between water permeability and compressive strength of CLSM mixtures was not found. Especially, Mixtures 2-L and 2-M, which contained wood ash W-2, showed higher permeability than other slurry mixtures. Permeability of Mixture 2-L even increased with age. This might be related to the significant swelling of Mixtures

2-L and 2-M (Table 4). In general, however, the permeability of the slurry mixtures decreased with age due to improved microstructure of the CLSM matrix resulting from continued pozzolanic and cementitious reactions of wood ash.

CONCLUSIONS

The following are the general conclusions of this investigation:

1. The three sources of wood ash, W1, W2, and W3, used in this investigation did not meet all of the requirements of ASTM C 618 for coal fly ash, especially loss on ignition. However, it is shown that they can be used as a main component in flowable slurry.

2. Wood ashes W-1 and W-3 caused expansive reactions in CLSM mixtures resulting in little or slight (average 1%) net shrinkage of CLSM. Wood ash W-2 caused either significant net swelling (15% for Mixture 2-L, and 21% for Mixture 2-M) or no shrinkage (Mixture 2-H) of CLSM.

3. The 91-day compressive strength of low-strength, medium-strength, and high-strength slurry mixtures was in the ranges of 0.38 to 0.97 MPa, 1.59 to 5.28 MPa, and 4.00 to 8.62 MPa, respectively. Overall, the slurry mixtures showed an average increase in strength of 150% (range: 24% to 446%) between the ages of 28 days and 91 days.

4. The wood ash from all three sources showed a noticeable increase in compressive strength of CLSM mixtures at later ages due to continuing pozzolanic and cementitious reactions.

5. In general, water permeability of the slurry mixtures decreased with age.

REFERENCES

1. National Council for Air and Stream Improvement (NCASI), "Solid Waste Management Practices in the U.S. Paper Industry - 1995," NCASI Technical Bulletin No. 793, NCASI, Research Triangle Park, NC, 1999.

2. ACI Committee 229, Controlled Low-Strength Materials (CLSM) (ACI 229R-99), ACI Manual of Concrete Practice, Part 1, American Concrete Institute, Farmington Hills, MI, pp. 229R-1 – 229R-15, 2000.

3. Naik, T.R. and Kraus, R.N., Use of Wood Ash for Structural Concrete and Flowable CLSM. Report No. CBU-2000-31, UWM Center for By-Products Utilization, College of Engineering and Applied Science, University of Wisconsin- Milwaukee, 121 pp., Oct. 2000.

4. Naik, T.R. and Singh, S.S., Development of Manufacturing Technology for Flowable Slurry Containing Foundry Sand and Fly Ash. Report No. CBU-1994-03, UWM Center for By-Products Utilization, College of Engineering and Applied Science, University of Wisconsin- Milwaukee, 95 pp., Aug. 1994.

5. Naik, T.R. and Singh, S.S., Permeability of Flowable Slurry Materials Containing Foundry Sand and Fly Ash. ASCE Journal of Geotechnical and Geoenvironmental Engineering, Vol. 123, No. 5, pp. 446-452, May 1997.

6. Naik, T.R. and Singh, S.S., Flowable Slurry Containing Foundry Sand. ASCE Journal of Materials in Civil Engineering, Vol. 9, No. 2, pp. 93-102, May 1997.

7. Naik, T.R. and Ramme, B.W., Low-Strength Concrete and Controlled Low-Strength Material (CLSM) Produced with Class F Fly Ash. ACI Special Publication SP-150, American Concrete Institute, Detroit, MI, pp. 1-14, June 1990.

8. Naik, T.R., Sohns, L.E., and Ramme, B.W., Controlled Low-Strength Material (CLSM) Produced with High-Lime Ash. Proceedings, ACAA 9th International Ash Utilization Symposium, Vol. 1, pp. 9-18, 1991.

9. Naik, T.R., Ramme, B.W., and Kolbeck, H.J., Filling Abandoned Underground Facilities with CLSM Fly Ash Slurry. ACI Concrete International, Vol. 12, No. 7, pp. 19-25, July 1990.

10. Ramme, B.W., Naik, T.R., and Kolbeck, H.J., Use of CLSM Fly Ash Slurry for Underground Facilities. Proceedings, Utilization of Industrial By-Products for Construction Materials, ASCE, New York, NY, pp. 41-51, 1993.

11. Krell, W.L., Flowable Fly Ash. ACI Concrete International, Vol. 11, No. 11, pp. 54-58, Nov. 1989.

12. Swaffer, K.M. and Price, M.R., Tunnel Saved by Fly Ash. Civil Engineering, ASCE, Vol. 57, No. 9, pp. 68-70, Sept. 1987.

13. Larson, R.L., Use of Controlled Low-Strength Materials in Iowa. ACI Concrete International, Vol. 10, No. 8, pp. 22-23, July 1988.

14. Larson, R.L., Sound Uses of CLSM in the Environment. ACI Concrete International, Vol. 12, No. 7, pp. 26-29, July 1990.

15. Fuston, J.J., Krell, W.C., and Zimmer, F.V., Flowable Fly Ash: A New Cement Stabilized Backfill. Civil Engineering, ASCE, Vol. 54, No. 3, pp. 48-51, March 1984.

16. Lai, C.I., Strength Characteristics of Flowable Mortars Containing Coal Ash. ACI Special Publication SP-132, Editor: V. M. Malhotra, American Concrete Institute, Detroit, MI, pp. 119-134, 1992.

17. Tikalsky, P.J., Smith, E., and Regan, R.W., Proportioning Spent Casting Sand in Controlled Low-Strength Materials. ACI Materials Journal, Vol. 95, No. 6, pp. 740-746, Nov.-Dec. 1998.

18. Tikalsky, P.J., Gaffney, M., and Regan, R.W., Properties of Controlled Low-Strength Material Containing Foundry Sand. ACI Materials Journal, Vol. 97, No. 6, pp. 698-702, Nov.-Dec. 2000.

19. Horiguchi, T., Okumura, H., and Saeki, N., Optimization of CLSM Mix Proportions with Combination of Clinker Ash and Fly Ash. Proceedings, Seventh CANMET/ACI International Conference on Fly Ash, Silica Fume, Slag, and Natural Pozzolans in Concrete, ACI Special Publication SP-199, Editor: V. M. Malhotra, American Concrete Institute, Farmington Hills, MI, pp. 307-324, 2001.

20. Horiguchi, T., Okumura, H., and Saeki, N., Durability of CLSM with Used Foundry Sand, Bottom Ash, and Fly Ash in Cold Regions. Proceedings, Fifth CANMET/ACI International Conference on Recent Advances in Concrete Technology, ACI Special Publication SP-200, Editor: V. M. Malhotra, American Concrete Institute, Farmington Hills, MI, pp. 325-348, 2001.

21. Naik, T.R., Kraus, R.N., and Singh, S.S., Use of Glass and Fly Ash in Manufacture of Controlled Low Strength Materials. Proceedings, Fifth CANMET/ACI International Conference on Recent Advances in Concrete Technology, ACI Special Publication SP-200, Editor: V. M. Malhotra, American Concrete Institute, Farmington Hills, MI, pp. 349-366, 2001.

22. Gassman, S.L., Pierce, C.E., and Schroeder, A.J., Effects of Prolonged Mixing and Retempering on Properties of Controlled Low-Strength Material (CLSM). ACI Materials Journal, Vol. 98, No. 2, pp. 194-199, Mar.-Apr. 2001.

Table 1 - Physical Properties of Wood Ash Used

Test	Source			ASTM C 618 Requirements		
	W1	W2	W3	Class C	Class F	Class N
Retained on 45 μm (No. 325) sieve (%)	23	60	90	≤ 34	≤ 34	≤ 34
Strength Activity Index with Cement (% of Control)						
3 days	88	38	102*			
7 days	84	39	83*	≥ 75	≥ 75	≥ 75
28 days	88	34	79*	≥ 75	≥ 75	≥ 75
Water Requirement (% of Control)	115	155	115*	≤ 105	≤ 105	≤ 115
Autoclave Expansion or Contraction (%)	0.2	0.5	-0.6*	≤ 0.8	≤ 0.8	≤ 0.8
Unit Weight (kg/m^3)	550	410	1380	—	—	—
Specific Gravity	2.26	2.41	2.60	—	—	—

* When material passing 150 μm (No. 100) sieve (25% of the wood ash W3) was used.

Table 2 - Chemical Composition of Wood Ash Used

Analysis Parameter	Source			ASTM C 618 Requirements		
	W1	W2	W3	Class C	Class F	Class N
Silicon Dioxide, SiO_2 (%)	32.4	13.0	50.7	—	—	—
Aluminum Oxide, Al_2O_3 (%)	17.1	7.8	8.2	—	—	—
Iron Oxide, Fe_2O_3 (%)	9.8	2.6	2.1	—	—	—
$SiO_2 + Al_2O_3 + Fe_2O_3$ (%)	59.3	23.4	61.0	≥ 50	≥ 70	≥ 70
Calcium Oxide, CaO (%)	3.5	13.7	19.6	—	—	—
Magnesium Oxide, MgO (%)	0.7	2.6	6.5	—	—	—
Titanium Oxide, TiO_2 (%)	0.7	0.5	1.2	—	—	—
Potassium Oxide, K_2O (%)	1.1	0.4	2.8	—	—	—
Sodium Oxide, Na_2O (%)	0.9	0.6	2.1	—	—	—
Sulfite, SO_3 (%)	2.2	0.9	0.1	≤ 5	≤ 5	≤ 4
LOI (750°C) (%)	31.6	58.1	6.7	≤ 6	≤ 6*	≤ 10
Moisture (%)	2.4	0.5	0.2	≤ 3	≤ 3	≤ 3
Available Alkali, Equivalent Na_2O (%)	0.9	0.4	0.8	≤ 1.5	≤ 1.5	≤ 1.5

* The use of Class F coal fly ash containing up to 12% loss on ignition may be approved by the user if either acceptable performance records or laboratory test results are made available.

Recycling Concrete and Other Materials

Table 3 - Mixture Proportions

	Low-Strength CLSM			Medium-Strength CLSM			High-Strength CLSM		
Mixture Number	1-L	2-L	3-L	1-M	2-M	3-M	1-H	2-H	3-H
Wood Ash Source	W1	W2	W3	W1	W2	W3	W1	W2	W3
WA/(C+WA) (%)	90	85	95	75	65	90	70	50	75
Cement, C (kg/m^3)	77	89	53	187	228	101	169	205	196
Wood Ash, WA (kg/m^3)	641	469	1187	611	400	1133	427	205	537
Water, W (kg/m^3)	626	635	481	602	596	510	418	430	359
W/(C+WA)	0.87	1.14	0.39	0.75	0.95	0.41	0.70	1.05	0.49
SSD Fine Aggregate (kg/m^3)	0	0	0	0	0	0	774	828	946
Flow/Spread (mm)	241	254	254	273	260	254	273	260	273
Air Content (%)	2.4	5.0	3.5	1.5	5.7	4.4	1.4	6.3	1.6
Air Temperature (°C)	22.2	22.2	22.2	25.6	25.6	25.6	25.6	26.1	25.6
Slurry Temperature (°C)	23.3	23.3	23.9	22.2	27.8	25.6	23.3	27.8	28.9
Unit Weight (kg/m^3)	1340	1190	1720	1390	1230	1760	1790	1670	2030

Table 4 – Change of Level of Top of CLSM with Reference to Top of Cylinder

Age (hour)	Level (mm)								
	1-L	2-L	3-L	1-M	2-M	3-M	1-H	2-H	3-H
0	0	0	0	0	0	0	0	0	0
1	-10	27	-10	-13	43	-5	-5	-10	-10
4	-13	38	-2	-10	51	-3	-3	-6	-6
8	-13	46	-2	-6	56	0	-2	-2	-2
24	-10	46	-2	-6	62	0	-2	0	0

Table 5 – Water Permeability of CLSM

Age (days)	Water Permeability ($\times 10^{-5}$ cm/sec)											
	Low-Strength CLSM				Medium-Strength CLSM				High-Strength CLSM			
	1-L	2-L	3-L	Avg.	1-M	2-M	3-M	Avg.	1-H	2-H	3-H	Avg.
28	10.1	5.5	1.5	5.7	0.6	51.3	2.0	18.0	1.1	0.2	11.9	4.4
91	8.0	12.3	1.3	7.2	0.1	35.0	0.2	11.8	0.4	0.1	3.8	1.5

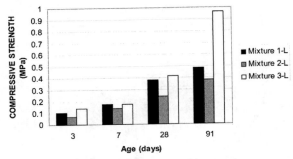

(Each result represents average of three tests.)

Fig. 1. Compressive strength of low-strength CLSM mixtures versus age.

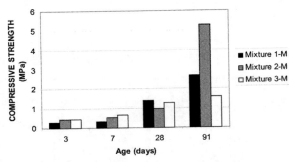

(Each result represents average of three tests.)

Fig. 2. Compressive strength of medium-strength CLSM mixtures versus age.

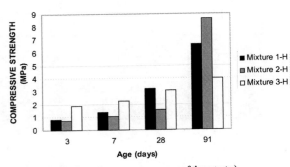

(Each result represents average of three tests.)

Fig. 3. Compressive strength of high-strength CLSM mixtures versus age.

SP-219—8

Protective System for Buried Infrastructure Using Recycled Tire Rubber-Filled Cement Mortars

by M. Nehdi and A. Khan

Synopsis: With population growth and urbanization, the space available for installation of civil infrastructure is rapidly decreasing. There is need for a more efficient use of underground space, which involves the construction of tunnels and other underground structures. Due to space constraints, many underground infrastructure projects in the future will be located in rock/soil with time-dependent behavior and/or under high overburden pressure. A deformable supporting system that can serve as a buffer layer for protecting tunnel linings and buried structures from time-dependant deformations of the excavated rock/soil will therefore be needed. This study investigates the possible use of cement mortars containing crumb tire rubber to develop a flexible interface material for such applications. The effects of the water/cement ratio (w/c) ratio, rubber content and particle size on the mechanical properties of the mortars were studied using uni-axial and tri-axial compression tests. A statistical factorial experimental was designed to obtain response surfaces for the parameters under study. The findings of this research suggest that cement mortars containing ground tire rubber have superior ductility and may be used to accommodate deformations around tunnel linings, pipelines, and other buried infrastructure.

Keywords: buried infrastructure; factorial experiments; recycling; stress-strain curve; tire rubber; tri-axial compression; tunnel linings; uni-axial compression

Moncef Nehdi is Associate Professor in the Department of Civil and Environmental Engineering, University of Western Ontario (UWO). He is a member of ACI committees 225 Hydraulic Cements, 236 Material Science of Concrete, 555 Recycling, and 803 Faculty Network Coordinating. His research interests include modeling the behavior of cement-based materials, recycling by-products in construction, and durability and repair of concrete infrastructure. Nehdi is recipient of the 2003 ACI Young Member Award for Professional Achievement.

Ashfaq Khan is currently a PhD candidate in the Department of Civil and Environmental Engineering at UWO. He obtained his MESc degree from the same Department in 2001. His research interests include applications of cellular concrete technology in tunneling projects and development of deformable cement-based materials.

INTRODUCTION

As the world's population is expected to exceed 9 billion people by year 2050 (Leon and Jane 1999), the demand for transportation infrastructure and lifeline systems such as wastewater collection and disposal networks will increase rapidly. In many already congested urban centers, this will lead to increased need for tunnels, pipelines and other underground structures that may be located in soft soils, in densely built areas, and in rocks with time-dependent deformations. Excavation in such field conditions relieves high *in-situ* stresses, providing an initiating mechanism for time-dependent deformations of soils and rocks. The construction of permanent rigid lining systems for underground structures restrains such movements, resulting in increased stresses acting on the lining system that may lead to its failure (Nehdi *et al.* 2002). This experimental study aims at developing a cost-effective flexible grouting material using ground recycled tire rubber that can act as a buffer layer between tunnel lining segments and the excavation line and accommodate excessive time-dependent deformations.

More than 281 million scrap tires (Rubber Manufacturers Association, 2002) are being discarded every year in the United States alone (Table 1), while another 500 million used tires are stockpiled throughout the country. These stockpiles are dangerous, not only because they pose a potential environmental threat, but they also constitute fire hazards and provide breeding grounds for mosquitoes (the recent West-Nile virus problem is a reminder that mosquitoes are no longer just a nuisance).

With growing restrictions on the disposal of scrap tires in landfills and associated environmental and safety concerns, there is an immediate need for identifying new uses and applications for recycled tire rubber. Tire rubber is being recycled to a limited degree for producing new tires. It is also being used successfully in asphalt mixtures. However, the initial cost of rubberized asphalt can be significantly higher than that of conventional asphalt and its long-term benefits are uncertain (Fedroff *et al.* 1996), while asphalt rubber programs take considerable time and resources to implement (Rubber Manufacturers Association, 2002).

Recycling Concrete and Other Materials

In addition, waste tires are being used as an alternative fuel in cement kilns, as feedstock for producing carbon black, and as reefs in marine environments (Nehdi and Khan 2001). Using tires as fuel is technically feasible but can be economically unattractive because of the high capital investment involved (O'Keefe 1984; Lee 1995). The use of rubber tires in the production of carbon black eliminates shredding and grinding costs but the carbon black from tire pyrolysis is more expensive, and gives lower quality carbon black than that from petroleum oils (Paul 1985). In addition, environmental concerns and public resistance have greatly impeded the option of incinerating waste tires. Although an economically attractive solution, the negative impact on the environment caused by tire incineration makes this alternative a compromise at best. Unfortunately, the generation of waste tires still exceeds its current uses. Table 2 shows the estimated total scrap tire use by various markets in the USA in 2001. About 77.6% of scrap tires are used.

The use of recycled tire rubber in cement-based materials has not received due attention because of the lower strengths associated with cementitious mixtures containing rubber granules (Nehdi and Khan 2001). However, recycled tire-rubber can be beneficial in developing deformable grouts that can be used for instance in tunnel linings as buffer layers between lining segments and the excavated tunnel face. Such flexible layers should have low strength and significant ductility in order to safely accommodate the time-dependent rock deformations and the corresponding stresses (Fig. 1). The scope of this paper is to investigate the stress-strain behavior of rubberized mortars under uni-axial and tri-axial compression including the effect of the w/c ratio, the rubber addition rate and particle size in order to optimize mixtures that are adequate for tunnel lining protection.

MATERIALS AND TEST METHODS

Ordinary portland cement (ASTM Type I), natural sand, tap water, and crumb tire rubber particles approximately 2 mm and 0.6 mm in average particle size (Fig. 2) were used to prepare 13 mortar mixtures. Tire rubber particles were added to the mortar mixtures at replacement levels of 0 to 100% per volume of sand according to the factorial experimental plan explained below. The mortar mixtures were prepared in a planetary mixer according to ASTM C305-82 guidelines (practice for mechanical mixing of hydraulic cement pastes and mortars of plastic consistency); their mixture proportions are summarized in Table 3. The rubber particles were added at the end of the standard mixing cycle and the mortar was mixed for one additional minute. For each mixture, six 3" x 6" (75 mm x 150 mm) test specimens were cast in plastic molds according to ASTM C192 guidelines (practice for making and curing concrete test specimens in laboratory). The moulded specimens were stored in a curing chamber at 100% relative humidity and temperature of 23±2 °C for 24 hours. After demoulding, specimens were capped with a sulfur solution according to ASTM C617-94 guidelines (practice for capping cylindrical concrete specimens). A uni-axial compression test (ASTM C39 – 96 standard test method for compressive strength of cylindrical concrete specimens) was performed on three of the six specimens for each batch at one day. The remaining specimens were stored in the curing chamber until testing for uni-axial compression at 28 days. An MTS machine with a loading capacity of 3000 kN was employed for the test (Fig. 3). The test was performed under deformation control at a rate of 1 mm/min. For each specimen, load versus deformation data was recorded using an automatic data logger. The data was then transferred to a spreadsheet and a stress-strain curve was obtained for each specimen. Secant modulus of elasticity at 50% stress (Neville 1995) was also obtained from the 1-day and 28-days stress-strain curves.

Four additional mortar mixtures were made to study the effect of the particle size of ground tire rubber on the stress-strain behavior of rubberized mortars under tri-axial stress conditions. Fine rubber particles having average particle size of 0.6 mm (Fig. 2) were incorporated into two mortar mixtures replacing 50% and 100% of sand by volume. For the remaining two mixtures, relatively coarser rubber particles having a mean particle size of 2 mm were used also replacing 50% and 100% of sand by volume. Table 2 summarizes the mixture proportions for these mortars. For each mortar mixture, three 2" x 4" (50 mm x 100 mm) cylindrical test specimens were cast, cured and capped according to ASTM guidelines as explained earlier. These specimens were then tested under tri-axial compression. ASTM C801-91 (standard test method for determining the mechanical properties of hardened concrete under tri-axial loads) was used to obtain the stress-strain behavior of each mixture under tri-axial loading. First, a test specimen was wrapped in a watertight rubber membrane and then placed in the tri-axial cell. An MTS loading machine having a maximum loading capacity of 245 kN was used to apply the vertical stress (major principal stress, σ_1). A hydrostatic/confining pressure of 0.3 MPa was applied and considered as the minor principal stress, σ_3 (Fig. 4). Axial load and axial strain were simultaneously monitored and recorded using an automatic data acquisition system.

STATISTICAL EXPERIMENTAL PLAN

In this study, the experiments were designed according to a factorial plan with two factors (input variables) at three levels (Fig. 5). Such factorial designs are widely used in experiments involving multiple factors where it is necessary to investigate not only the effects of the variables but also those of their interactions. This approach also allows limiting the number of experiments while first and second order models could be used to fit the data and predict the response of other experimental points located within the experimental domain (Myers and Montgomery 1995). The sequence in which the experimental points were investigated was randomized to avoid any statistical significance of a blocking effect.

A second order model as given below in equation (1) was found suitable to fit the data obtained.

$$Y = \beta_0 + \beta_1 X_1 + \beta_2 X_2 + \beta_{11} X_1^2 + \beta_{22} X_2^2 + \beta_{12} X_1 X_2 + \varepsilon \quad (1)$$

Where:
- Y: response of experiment
- X_1 and X_2: experimental variables
- β_i and β_{ij}: coefficients of the model
- ε: random error term

Recycling Concrete and Other Materials

In the present experimental plan, the two factors (experimental variables) were the water/cement ratio (w/c) and the proportion of recycled tire rubber (RR) replacing equal volumes of sand. All other controllable parameters were kept constant. For all the designed mortars, response surfaces for compressive strength and maximum strain (strain at which the MTS load dropped to near zero) at 1-day and 28-days were obtained. The data corresponding to the various responses that resulted from the designed experimental plan were analyzed and plotted using a statistical software package (Design Expert, 1998).

RESULTS AND DISCUSSION

Uni-Axial Compression Tests

Figure 6 illustrates response surfaces of uni-axial compressive strength results at 1-day and 28-days for various rubber-filled mortars. It is shown that the compressive strength of rubber included mortars decreased with increased RR addition and with higher w/c. However, increasing the RR addition appeared to be the dominant factor in reducing the uni-axial compressive strength of these mortars.

Figure 7 illustrates the stress-strain curves under uni-axial compression of three of the rubber-filled mortar mixtures containing 0%, 50% and 100% of ground tire rubber, respectively but having equal w/c (0.5). For the mixture containing 50% RR by volume of sand, the compressive strength was reduced by 50% at 1-day and 75% at 28-days compared to corresponding strengths of the control mixture without RR. However, both the 1-day and 28-days compressive strengths of the mixture containing 100% RR by volume of sand were reduced by around 85% compared to corresponding strength values of the control mixture without RR.

Figure 8 shows the response surfaces of maximum strain (strain at which the MTS load dropped to near zero) at 1-day and 28-days for various rubber-filled mortars. The maximum strain values at 1-day and 28-days for cementitious mortars containing recycled tire rubber significantly increased as the w/c and proportions of RR increased. Figure 7 in which the stress-strain curves of three of these mortar mixtures are compared further illustrates this effect. It is shown that the descending portions of the stress-strain curves for the specimens containing 50% RR indicated some ductile behavior with limited strain softening. For such specimens, the maximum strain at 1-day and 28-days increased by 100% compared to the maximum strains observed for the control specimens without RR (Fig. 7a and 7b). However, specimens containing 100% RR demonstrated more significant ductility; their maximum strain at 1-day and 28 days increased by 150% compared to corresponding values of the control specimens without RR (Fig. 7a and 7b).

It is argued that the addition of ground tire rubber decreased the elastic modulus of the rubber-filled mortars and thus enhanced their flexibility. The secant modulus of elasticity at 50% stress was obtained both at 1-day and 28-days from the stress-strain curves. It was found that the secant modulus of elasticity for the mortar specimens containing 50% RR by volume of sand was reduced to 25% at 1-day and to 45% at 28-days of that of the control specimens without RR. For the specimens containing 100% RR by volume of sand, the secant modulus of elasticity was reduced to 12% at 1-day and to 20% at 28-days of that of the control specimens without RR. Experimental results of Schimizze et al. (1994) and Khatib and Bayomy (1999) also showed a comparable trend.

Tri-axial Compression Tests

The tri-axial compression test can better simulate three-dimensional stress conditions and anisotropic behavior of rock/soil mass around tunnels and underground structures than the uni-axial compression test. Figure 9a compares the stress-strain curves under tri-axial compression of three mortar mixtures containing 0% rubber, 50% coarse recycled rubber (CRR) and 50% fine recycled rubber (FRR), respectively by volume of sand. Both mixtures containing 50% CRR and 50% FRR demonstrated lower compressive strengths but higher ductility than that of the control mixture without rubber addition. However, the mixture containing 50% FRR demonstrated slightly higher compressive strength and lower maximum strain (strain at which the deviator stress dropped to near zero) when compared to the mixture containing relatively coarser rubber particles. The mixture containing CRR showed a maximum strain of 6% compared to 5.7% for the mixture with FRR and 3.8% for the control mixture.

Figure 9b compares the stress-strain curves of three mortar mixtures containing 0% RR (control mixture), 100% CRR and 100% FRR, respectively replacing an equal volume of sand. The mixture containing 100% CRR demonstrated better ductile behavior with the maximum strain reaching about 8%. Such a ductile behavior could be utilized in geotechnical engineering applications such as the development of flexible interface buffer materials to protect tunnel linings and underground structures from lateral deformations of rock/soil. The mixture containing FRR exhibited relatively lower plastic deformation.

When a flexible buffer layer is used around underground structures, lateral stresses due to the surrounding rock/soil can develop in the lining system up to the crushing strength of the buffer layer. The stresses on the lining system will not exceed this level until the crushing of the flexible buffer layer can no longer accommodate the lateral deformations of the surrounding rock/soil. If the buffer layer has adequate thickness and a low crushing strength, it may safely accommodate the lateral stresses due to surrounding rock/soil for the entire design life of the structure. Because the crushing strength of the buffer layer should be low, the mixture proportions of the rubber-included mortars in Table 3 and Table 4 can be altered so that the cementitious material incorporates high-volumes (50-70%) of recycled by-products such as fly ash and slag. Combined with recycled tire rubber, the use of such a material can provide unique economic and environmental benefits.

Mode of Failure

The test specimens of rubber-filled mortars did not exhibit brittle failure when loaded under uni-axial or tri-axial compression. The observed failure was rather gradual of a shearing mode (Fig. 10). The specimens continued to sustain load after peak load and did not fracture into pieces. Upon release of the load, the specimens rebounded back to near their original shape. Such behavior for rubber-filled concrete mixtures was also reported by Biel and Lee (1996).

It was argued (Eldin and Senouci 1993) that since cement paste is much weaker in tension than in compression, rubber-included concrete specimens containing coarse tire chips would start failing in tension. The generated tensile stress concentrations at the top and bottom of the rubber aggregates (Fig. 11a) result in many tensile micro-cracks that form along the tested specimen (Fig. 11b). These micro-cracks will rapidly propagate in the cement paste until they encounter a rubber particle. Because of their ability to withstand large tensile deformations, the rubber particles will act as springs (Fig. 11c), delaying the widening of cracks and preventing full disintegration of the concrete mass. The continuous application of the compressive load will cause generation of more cracks and widening of existing ones. During this process, the failing specimen is capable of absorbing significant plastic energy and withstanding large deformations without full disintegration. This process will continue until the stresses overcome the bond between the cement paste and the rubber particles.

It is believed that tensile failure of the mortar around rubber particles is a localized occurrence. These occurrences and the weakening of the bond between the cement paste and rubber aggregates sufficiently weaken the overall grout matrix such that it fails in shear, but at a greatly reduced strength compared to a control specimen without rubber addition. Conversely, typical failure of the control specimens without rubber resulted in explosive conical fractures of the specimens, leaving them in several pieces.

CONCLUSIONS

Currently, the tunnel engineering industry is focusing on mining large diameter tunnels with TBMs (tunnel boring machines) and supporting the excavation line with single-pass segmental concrete lining systems. However, in order to be technically feasible, the design of single-pass systems requires the development of a compressible grout annulus between the tunnel lining and the rock mass to reduce and control the ground induced lateral pressures and stresses in the lining system. This paper investigated the flexibility of rubber-filled mortar mixtures in order to develop a deformable interfacial material for tunnel linings and underground structures. The factorial experimental plan used for this study provided response surfaces for the various parameters investigated based on 13 mortar mixtures. Such response surfaces offer a simple visual tool to compare the performance of rubber-filled mortars. Various specimens of rubber-filled mortars were tested under uni-axial and tri-axial compression and the stress-strain behavior of each mixture was studied. The following conclusions can be warranted based on this study:

i. Using recycled tire rubber as partial replacement for sand in cementitious mortars resulted in a reduction of compressive strength of the mortar mixtures. The compressive strength was reduced by around 85% both at 1-day and 28-days for specimens containing 100% replacement of sand by RR. For specimens containing 50% replacement of sand by RR, the compressive strength was reduced by 50% at 1-day and 75% at 28-days compared to that of the control mixture without RR.

ii. Stress-strain curves of mixtures containing 50% RR replacing equal volume of sand demonstrated significant plastic deformation. However, mixtures containing 100% rubber by volume of sand exhibited higher plastic deformation at the expense of larger reductions in compressive strength.

iii. Failure of rubber-filled specimens was gradual and in a shear mode, resulting in diagonal fracture planes that traversed the cylindrical specimens. It is believed that tensile failure of the mortar around the rubber particles is a localized occurrence. These occurrences sufficiently weaken the overall grout matrix such that it fails in shear, but at a significantly reduced strength.

iv. Increasing the RR addition resulted in a decrease in elastic modulus of rubber-filled mortars, which indicates higher flexibility, and may be viewed as a positive gain in rubber-filled mortar mixtures that could be used in developing flexible grouting systems for tunnel linings with rock/soil induced stresses. Buffer layers of such flexible mortars can be placed around lining segments of underground structures to absorb rock/soil squeeze and other lateral deformations and stresses.

v. Further larger scale tests including other aspects such as fire resistance should be carried out before field implementation of rubber-filler mortars in tunnel linings and other underground infrastructure.

ACKNOWLEDGEMENT

The authors are grateful to the Charles A. and Anne Morrow Lindbergh Foundation, Anoka, MN, USA for the financial support of this research project.

REFERENCES

1. Biel, T. D., and Lee, H. (Nov. 1996), "Magnesium oxychloride cement concrete with recycled tire rubber", Trans. Res. Rec., No. 1561, pp. 6-12.

2. Design Expert, Version 5.0.3 (1998), STAT-EASE-Inc., 2021 East Hennepin Ave., Suite 191, Minneapolis, MN 55413.

3. Eldin, N. N., and Senouci, A. B. (1993), " Rubber-tire particles as concrete aggregates", J. Mat. Civ. Engg, ASCE, Vol. 5, No. 4, pp. 478–496.

4. Epps, J. A. (1994), "Uses of recycled rubber tires in highways", Synthesis of Highway Practice 198, TRB, Washington, D.C., 162 p.

5. Fedroff, D., Ahmad, S., and Savas, B. Z. (Sep. 1996), " Mechanical properties of concrete with ground waste tire rubber", Trans. Res. Rec., No. 1532, pp. 66–72.

6. Khatib, Z. K., and Bayomy, F. M. (Aug. 1999), "Rubberized portland cement concrete", J. Mat. Civ. Engg., pp. 206–213.

7. Lee, B. (1995), Presentation at ACI Spring Convention, ACI, Salt Lake City, Utah.

8. Leon, F. B. and Jane, T. B. (1999), World Population: Challenges for 21st Century, Seven Locks Press – Santa Ana, CA 92799, USA, p. 16.

9. Myers, R. H. and Montgomery, D. C. (1995), Response Surface Methodology: Process and Product Optimization Using Designed Experiments, Wiley – Inter-Science Publication, pp. 1-15.

10. Nehdi, M. and Khan, A. (June 2001), "Cementitious composites containing recycled tire rubber: an overview of engineering properties and potential applications", ASTM Journal of Cement, Concrete and Aggregates, Vol. 23, No. 1, pp. 3-10.

11. Nehdi, M., Khan, A., and Lo, K.Y. (2002), "Development of a deformable protective system for underground infrastructure using cellular concrete", ACI Materials Journal, Vol. 99, No. 5, September/October 2002, pp. 490-498.

12. Neville, A. M. (1995), Properties of Concrete, 4th ed., Longman Group Limited, Essex CM20 2JE, England, pp. 412-414.

13. O'Keefe, W. (Oct. 1984), Power, Vol. 128, No. 10, p. 115.

14. Paul, J. (1985), Encycl. Poly. Sci. Engg., Vol. 14, pp. 787-802.

15. Schimizze, R., Nelson, J., Amirkhanian, S., and Murden, J. (1994), " Use of waste rubber in light-duty concrete pavements", In Proc. ASCE 3rd Mat. Engg. Conf., Infrastructure: New Materials and Methods of Repair, pp. 367-374.

16. SHRP, Strategic Highway Research Program-Report. (July 1992), Funding Available for Materials Recycling Research.

17. US Scrap Tire Markets 2001 (December 2002), Rubber Manufacturers Association, 1400 K Street, NW, Washington DC 20005, 34 p.

Table 1 – Statistics for scrap tires

Country	Number of scrap tires (Millions/year)
USA-1996-1998-2001	266-270-281
Japan-1996	102
France-1996	44.3
Germany-1996	28.2
Great Britain-1996	23.4
Canada-1996	20.0
Australia-1996	17.0
Italy-1996	12.1
Country	Number of scrap tires stockpiled
USA–1996-1998	800-500
Europe-1990	161

Table 2 – Estimated total US scrap tire use markets in 2001 (after Rubber Manufacturers Association)

Market	Millions of tires used
Tire derived fuel	
Cement kilns	53
Pulp/paper mills	19
Tires dedicated to energy	14
Electric utilities	18
Industrial boilers	11
Total fuel use	115
Products	41
Ground rubber	33
Cut punched/stamped	8
Civil engineering	40
Miscellaneous/agriculture	7
Export	15
TOTAL USE	218
TOTAL GENERATION	281
Use as % of total generation	77.6%

Table 3- Mixture proportions for mortars tested under uni-axial compression

Batch #	Cement (kg/m^3)	Water (kg/m^3)	Rubber (kg/m^3)	Sand (kg/m^3)	W/C Ratio	Unit Weight (kg/m^3)
1	440	265	0	1535	0.6	2240
2	470	190	135	880	0.4	1675
3	440	220	135	835	0.5	1630
4	450	225	275	0	0.5	950
5	570	230	0	1535	0.4	2335
6	440	220	135	835	0.5	1630
7	440	220	135	835	0.5	1630
8	440	220	135	835	0.5	1630
9	400	240	275	0	0.6	915
10	520	205	275	0	0.4	1000
11	490	255	0	1535	0.5	2280
12	440	220	135	835	0.5	1630
13	390	235	135	835	0.6	1595

Table 4- Mixture proportions for mortars tested under tri-axial compression

Batch #	Cement (kg/m^3)	Water (kg/m^3)	Rubber (kg/m^3)	Sand (kg/m^3)	W/C Ratio	Unit Weight (kg/m^3)
1	490	255	0	1535	0.5	2280
2	450	225	275	0	0.5	950
3	440	220	135	835	0.5	1630
4	540	275	295	0	0.5	1110
5	450	225	160	850	0.5	1670

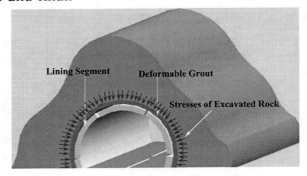

Fig. 1 Stresses of excavated rock acting on deformable mortar and ridgid lining segments.

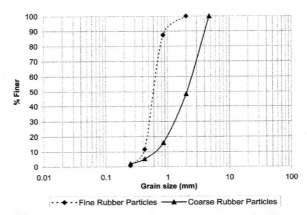

Fig. 2 Particle size distribution for fine and coarse recycled tire rubber particles.

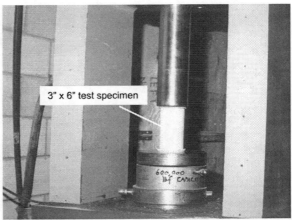

Fig. 3 MTS machine applying uni-axial compression on a 3" x 6" cylindrical rubber-filled mortar specimen.

Recycling Concrete and Other Materials 111

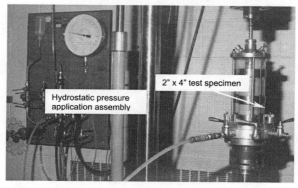

Fig. 4 2" x 4" rubber-filled mortar specimen under tri-axial compression.

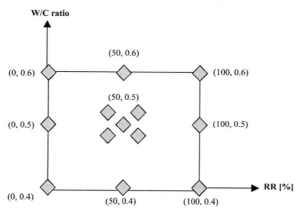

Fig. 5 Illustration of the factorial experimental plan used (RR = recycled rubber).

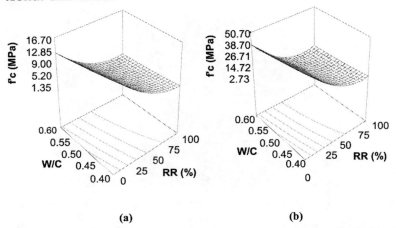

Fig. 6 Response surfaces of (a) 1-day f'c and (b) 28-day f'c for rubberized cement mortars.

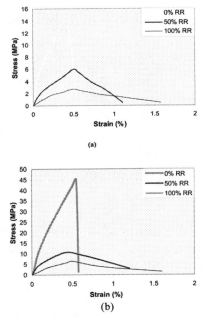

Fig. 7 Stress-strain curves for rubberized mortars under uni-axial compression at (a) 1-day and (b) 28-days.

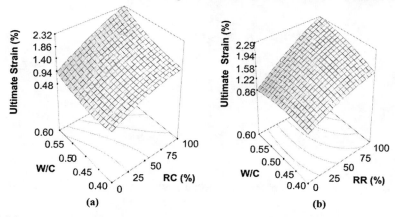

Fig. 8 Response surface of (a) 1-day maximum strain and (b) 28-day maximum strain for rubberized cement mortars

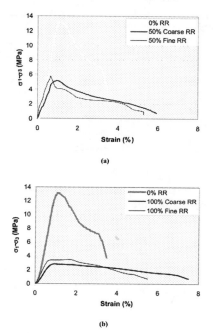

Fig. 9 Stress-strain curves under tri-axial compression at 1-day for mortars containing (a) 50% and (b) 100% of coarse and fine RR versus a control mixture with RR.

Fig. 10 Mode of failure of rubber-filled mortar specimens after (a) uni-axial compression test (b) tri-axial compression test

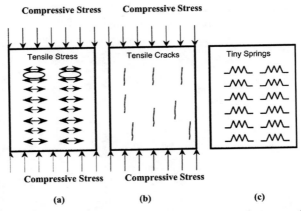

Fig. 11 Modeling the behavior of rubber-filled test specimens under uni-axial compression (after Eldin and Senouci 1993).

SP-219—9

The Use of Crushed Limestone Dust in Production of Self-Consolidating Concrete (SCC)

by C. Shi, Y. Wu, and C. Riefler

<u>Synopsis:</u> Crushed limestone dust is a waste material from the production of concrete aggregate by crushing quarried limestone rocks. The dust is usually less tan 1% of the aggregate production. Although it is coarser than common cementing materials such as as Portland cement, coal fly ash and ground blast furnace slag, it is fine enough to cause many problems during materials handling and disposal. Laboratory results have indicated that crushed limestone dust can be used to produce self-consolidating concrete (SCC) with properties similar to those of SCC containing coal fly ash. . Due to the differences in morphologies and particle size distribution, the mix design has to be modified when crushed stone dust instead of fly ash or ground blast furnace slag is used. Fresh SCC mixtures containing limestone dust loses its flowability and sets faster than the mixtures containing fly ash due to the acceleration of the hydration of Portland cement by the limestone powder. SCC containing limestone dust exhibited strengths similar to that containing fly ash during the first seven days, but the former exhibited lower strength than the latter at 28 and 90 days due to the contributions from the pozzolanic reactions between coal fly ash and lime released from the hydration of Portland cement. The former also have lower autogenous and drying shrinkages than the latter.

<u>Keywords:</u> autogenous shrinkage; coal fly ash; drying shrinkage; flowability; limestone dust; segregation; self-consolidating concrete; strength

ACI member **Caijun Shi** is president of CJS Technology Inc., Ontario, Canada. He received his Ph.D. from the University of Calgary, Canada in concrete materials science in 1993. He is a member of ACI committees 232, 233, 236, 523 and 555. His research interests include design and testing of concrete materials, use of industrial by-products and recycled materials in concrete, and stabilization/solidification of hazardous wastes with cements.

Yanzhong Wu is a Research Engineer of CJS Technology Inc., Ontario, Canada. He received his Ph.D. from the University of Dundee, Scotland in concrete technology in 1995. His research interests include design and testing of cement and concrete materials

Chris Riefler is president of Riefler Development L.L.C., Hamburg, New York. He worked as a sales engineer, plant manager and vice president of operations for more than 20 years in Riefler Concrete Products, L.L.C. in Hamburg, New York. He has extensive experience with the manufacture and quality control of ready mix concrete and various concrete products.

INTRODUCTION

SCC has many advantages over conventional concrete: (1) eliminating the need for vibration; (2) decreasing the construction time and labor cost; (3) reducing the noise pollution; (4) improving the filling capacity of highly congested structural members; (5) improving the interfacial transitional zone between cement paste and aggregate or reinforcement; (6) decreasing the permeability and improving durability of concrete, and (7) facilitating constructibility and ensuring good structural performance. SCC has attracted more and more attention world-wide since its discovery, and new applications for SCC are being explored because of its advantages [1-5].

Several design procedures based on scientific theories or empirical experiences have been proposed for SCC [6-11]. In general, these procedures fall into the following two categories: (1) combination of superplasticizer and high content of mineral powders, and (2) combination of superplasticizer and viscosity-modifying admixture (VMA) with or without defoaming agent. Shi et al [12] compared the properties of SCCs designed using the two approaches. The use of these chemical admixtures and/or mineral powders is to achieve required flowability, passing ability and good segregation resistance [13-16]. Passing ability refers to the ability of SCC to flow through tight openings such as spaces between steel reinforcing bars without segregation or blocking. Good segregation resistance means that the aggregates distribute uniformly at all locations and at all levels. Aggregate segregation will have effects on the deformability, blocking around reinforcement, high drying shrinkage, and compressive strength of the concrete mixtures.

The commonly used powders include coal fly ash, calcined clays, ground glass and blast furnace slag. Different powders can show very significant effects on the requirement of superplasticizer, deformability, filling capacity and strength of concrete [17-21]. Fly ashes have spherical particles and may decrease the water requirement for given flowability and superplasticizer dosage or decrease superplasticizer dosage for a given water content and flowability. The other powder materials have irregular particles and their effects on water requirement or flowability are dependent on their particle size and shape. Fly ash, calcined clay, ground glasses and blast furnace slag have pozzolanic or cementitious properties. In most cases, they may retard the setting and decrease early strengths, but can increase later strengths and improve the durability of concrete.

The addition of limestone powder into cement can affect the hydration of cement, the rheological properties of fresh cement pastes and concrete mixtures and properties of hardened concrete [22-29]. Crushed limestone dust is a waste material from the production of concrete aggregate by crushing quarried rocks. The dust is usually less than 1% of the aggregate production. Previous work has indicated that crushed granite dust can be used as a powder for the production of SCC, but requires a higher superplasticizer dosage than limestone powder to achieve similar yield stress and other rheological properties [18]. The use of crushed stone dust in SCC can decrease the materials cost of SCC, eliminate the dust disposal cost and reduces environmental pollutions. The purpose of this study is to investigate the properties of SCC using crushed limestone dust as a filler in comparison with that using ASTM Class F fly ash, which is widely used in normal concrete and SCC.

RESEARCH SIGNIFICNCE

Crushed limestone dust is a waste material from the production of concrete aggregate by crushing quarried rocks. This research was undertaken to investigate the properties of SCC containing crushed limestone dust. Although it consists mainly of elongated particles and is much coarser than coal fly ash or Portland cement, it can be used satisfactorily for production of SCC. SCC containing shows less shrinkage than SCC containing coal fly ash.

EXPERIMENTATION

Raw materials

The stone dust is from a limestone crushing operation collected by a dust collector. Its particle size distribution is shown in Fig. 1. SEM observations indicated that the limestone dust consists mainly of elongated irregular particles.

A typical commercial Type I Portland cement, which complies with the requirements of Specification ASTM C 150 was used as the testing cement. A commercial ASTM Class F coal fly ash was used as a reference. The chemical compositions of Portland cement and fly ash are shown in Table 1. The physical properties of the cement and fly ash, as supplied by the manufacturers, are summarized in Table 2. The superplasticizer used in this study was a polycarboxylate derivative-based superplasticizer.

Crushed limestone with a maximum size of 15 mm was used as the coarse aggregate. Natural fine sand was used as fine aggregate. The gradations of the two aggregates are shown in Fig. 2. The fineness modulus of the sand is 2.89.

Mixture design

In this study, the SCCs were designed by partial replacement of aggregates with limestone dust to achieve high flowability and passing ability without showing segregation. The fine to coarse aggregate ratio was optimized to achieve the least void content and to minimize the required paste content. Two SCC mixtures, one using limestone dust as powder material (SCC-SD) and the other one using class F fly ash (SCC-FA) are summarized in Table 3.

Testing of fresh SCC mixtures

Concrete mixtures were mixed in a high-speed shear mixer. Air content, density, slump cone flowability, and L-box flowability of the concrete mixtures were measured. Air content and density of fresh concrete mixture were determined following ASTM C231 and C138 respectively.

Slump cone flowability of the concrete mixtures was measured using a regular slump cone. The slump cone was filled with concrete mixtures without rodding or tamping, and then lifted up vertically. The diameter of the mixture after unconfined lateral spread was recorded as the flowability of the mixture. Four measurements were taken and an average of the four measurements are presented. The slump cone flowability of the SCCs was also measured at 5, 30 and 60 minutes after the addition of mixing water. Between measurements, the mixtures were stored in a bucket and covered with a damp cloth.

L-box test can assess flowability, filling ability and passing ability of SCC. Filling ability refers to ability of SCC to flow into and fill completely all spaces within the formwork, under its own weigh. L-box, as shown in Fig.3, consists of a vertical and a horizontal section in an "L" shape, and separated by a sliding gate. A set of vertical reinforcement bars (3xØ12mm) is fitted in front of the gate. The vertical section is filled with a concrete mixture without rodding or tamping. After the sliding gate is lifted, concrete will flow into the horizontal section. The time for concrete to reach the end of the horizontal section is recorded. The height of the concrete in the vertical section (H1) and the height of concrete at end of the horizontal section (H2) are measured. Any blockage of concrete behind the reinforcing bars can be visually observed.

Penetration resistance and setting times

The development of penetration resistance of the two SCCs was measured per ASTM C403. The measured results were fitted using equations as recommended by ASTM C403. Initial setting time refers to the time with a penetration resistance of 3.5 MPa, and final setting time corresponds to a penetration resistance of 28 MPa.

Specimen preparation and testing of hardened SCCs

For each batch, twenty-four 100x200 mm cylinders were cast for compressive strength testing and six 75x75x285 mm prisms were cast for autogenous and drying shrinkage testing. SCC mixtures were cast into molds in one layer without any compaction or vibration. After casting, the molded specimens were taken to a fog room at 23±2°C and 100% relative humidity and covered with a plastic sheet.

For strength testing, cylindrical specimens were demolded 24 hours after casting and placed back into the fog room for continued curing. Three cylinders were taken out for compression tests at each testing age. The reported result is an average of three specimens.

For autogenous shrinkage testing, two prismatic specimens from each batch were sealed and measured for the first reading after the demolding. These sealed specimens were placed in a room at 23±2°C and measured for lengths at different ages. Another two prisms from each batch were placed into a conditioned chamber at 23±2°C and 50±5% relative humidity after demolding for drying shrinkage testing. The last two prisms from each batch were cured for another 6 days in the fog room after demolding then transferred to the conditioned chamber for drying shrinkage testing. The moisture losses of these specimens during drying shrinkage were also measured. The reported shrinkage or mass loss is an average of two specimens.

RESULTS AND DISCUSSIONS

Properties of fresh concrete

Properties of the two fresh concretes are summarized in Table 4. Although crushed limestone dust consists mainly of irregular and elongated particles [18], fresh SCC-SD exhibited similar rheological behavior as fresh SCC-FA. Neto and Campiteli [23] found that a reduction of the yield value with the increase of limestone dust content, and a slight increase of plastic viscocity after a certain relation of fineness to limestone content. Both SCC mixtures had similar air content, 2.9% for SCC-SD mixture and 2.2% for SCC-FA mixture, which are typical values for non-air-entrained concrete. During the slump cone flow test, it was observed that the designed SCCs exhibited a very homogenous mixture and did not show observable indication of bleeding or segregation. SCC-FA and SCC-SD had H2/H1 ratios of 76% and 79% respectively from L-box testing. Usually, an H2/H1 ratio of greater than 80% is required for SCC. However, the L-box used in this study has a horizontal length of 800 mm instead of 700 mm for most cases. Thus, the H2/H1 ratios for the two SCCs should still be regarded as satisfactory. SCC-FA also flowed faster than SCC-SD during the L-box testing, which was consistent with the slump cone flowability and H2/H1 ratio.

The effect of fly ash on properties of fresh concrete is dependent on the characteristics of the fly ash. Carbon content in fly ash has an adverse effect on workability; thus, variations in carbon content may lead to erratic behavior with respect to workability and air entrainment [30]. A concrete containing fly ash is cohesive and has a reduced bleeding capacity. It is not the purpose of this study to investigate how coal fly ashes affect the workability of SCC, but the test results indicated that the SCC containing coal fly ash performed satisfactorily.

Fig. 4 shows the change of slump cone flowability of the two SCCs with time up to 60 minutes. The flowability of SCC-FA decreased almost linearly with time. However the flowability of SCC-SD deceased faster in the beginning but then slowed down. Flowability of the SCC containing limestone dust decreased more quickly. This could be resulted from two factors: 1) physically, the fine stone dust particles act as nucleation sites accelerating the hydration of cement, and 2) chemically, limestone dust reacts with C_3A in the cement to form carboaluminates.

Setting of SCC mixtures

The development of penetration resistance of SCC concretes is shown in Fig.5. It can be seen that SCC-SD develops penetration resistance faster than SCC-FA. Fly ash in concrete has a retarding effect, typically of about one hour, probably caused by the release of SO_3^{2-} present on the surface of some fly ash particles [31]. On the other hand, the presence of limestone powder can accelerate the hydration of cement [24]. The setting times for the two concrete mixtures are summarized in Table 5. The initial and final setting times of SCC-FA are approximately 45 minutes and 80 minutes longer than the SCC-SD.

Compressive strength of hardened SCCs

Strength development of the two SCCs is shown in Fig. 6. During the first 7 days, the two concretes had almost the same strength, but the rate of strength gain was significantly different thereafter. The 28-day and 90-day strengths for SCC-SD were 40 MPa and 43 MPa respectively. The 28-day and 90-day strengths for SCC-FA were 48 MPa and 59 MPa respectively. As mentioned above, the presence of limestone powder accelerates the hydration of cement, thus, it can be expected that its presence can also increase the early strength of the concrete through its chemical reaction with cement [29]. It has also been noticed that SCC using limestone powder as a filler exhibited higher strengths than normal concrete for a given cement concrete [25]. This can be contributed to both the acceleration of hydration of the cement by the limestone powder and the homogenous structure of the SCC mixtures.

The addition of fly ash into concrete usually also slightly increases the strength of concrete at early ages. However, as the pozzolanic reactions between fly ash and $Ca(OH)_2$ released from the hydration of Portland cement starts, it will increase the strength of the concrete significantly. This explains why SCC-FA and SCC-SD showed similar strength at early ages, but the former had significantly higher strengths at later ages.

Autogenous shrinkage

Autogenous shrinkage refers to the shrinkage resulting from the hydration reactions of cements. The autogenous shrinkage of the two SCCs after one day is shown in Fig.7. It can be seen that, during the first 5 days, both SCCSs did not exhibit measurable autogenous shrinkage. After 5 days, the autogenous shrinkage of both SCCs increased with time. After around 10 days, the gain rate of autogenous shrinkage of both SCCs started to decrease with time, but SCC-FA still increased much faster than SCC-SD. Therefore the difference in autogenous shrinkage between SCC-SD and SCC-FA increased with age. At 56 days, the autogenous shrinkages of SCC-SD and SCC-FA were 0.004% and 0.01% respectively. As discussed above, the difference at late ages between the two SCCs can be attributed to the pozzolanic reactions between fly ash and $Ca(OH)_2$ released from the hydration of Portland cement.

Drying shrinkage

Drying shrinkage and moisture loss of the two SCCs after one day of moisture curing are plotted in Figs. 8 and 9. Although SCC-SD and SCC-FA showed the same moisture loss, the former showed a much lower drying shrinkage than the latter. Many mechanisms for shrinkage, including surface tension, capillary tension and disjoining pressure, have been proposed for conventional hardened cement and concrete materials [32]. It is generally agreed that several mechanisms co-exist. The capillary tension due to the loss of water in capillary pores is probably the most important factor in determining the drying shrinkage under the current testing conditions. However, capillary tension is determined by the critical pore size rather than the total volume of the pore from which water is lost [33]. The smaller the pore size is, the larger the capillary tension or drying shrinkage is. Thus, the larger drying shrinkage of SCC-FA can be attributed to a smaller critical pore size than that in SCC-SD.

Adams and Race [26] noticed that the addition of interground or blended limestone in Portland cement could adversely affect the drying shrinkage when no other changes are made to the cement. It was noticed that the drying shrinkage of SCC with the addition of fly ash or limestone powder showed less drying shrinkage than normal concrete [27, 28]. The reason is that, for the same water content, increasing powder volume and thus lowering the w/p will reduce drying shrinkage.

The drying shrinkage and moisture losses of the two SCCs after seven days of moisture curing are plotted in Figs. 10 and 11. Increasing the initial moist curing period from one to seven days did not show an effect on the drying shrinkage of SCC-SD, by comparing Figs. 8 and 10. However, extended initial moisture curing showed a great effect on the drying shrinkage of SCC-FA. The longer the initial moisture curing period is, the lower the drying shrinkage is. The 28- and 50-day drying shrinkages were reduced approximately by 30% when the initial moisture curing period was increased from 1 day to 7 days.

Comparing the results in Fig.11 with those in Fig.7, it can be seen that extended initial moisture curing decreased the moisture losses of the two SCCs during drying tests. However, SCC-SD still exhibited higher moisture loss than SCC-FA. As discussed above, it is the critical pore size rather than the total pore volume of the concrete that controls the drying shrinkage of the concrete during drying.

CONCLUSIONS

The following conclusions can be drawn from this study:

Crushed limestone dust collected in a dust collector from crushing operations consists mainly of elongated particles and is much coarser than coal fly ash or Portland cement. However, it can be used satisfactorily for production of SCC.

The properties of fresh SCC mixture containing limestone dust are very similar to those of the mixture containing fly ash. However, the former loses its flowability faster.

SCC containing limestone dust set faster than the mixture containing coal fly ash. This can be attributed to the fact that limestone powder accelerates the hydration of cement, while fly ash retards the setting of concrete.

The two SCCs exhibited similar strengths during the first seven days. However, the SCC containing limestone dust gained less strength than the SCC containing coal fly ash due to the contributions from the pozzolanic reactions between coal fly ash and lime released from the hydration of Portland cement.

SCC containing limestone dust exhibited lower autogenous and drying shrinkage than the SCC containing fly ash due to the pozzolanic reactions between coal fly ash and lime released from the hydration of Portland cement. Moisture losses of the former were higher than those of the latter during the drying shrinkage testing.

REFERENCES

1. Okamura, H. and Ouchi, M., "Self-Compacting Concrete. Development, Present and Future," Proceedings of the 1st RILEM Symposium on Self-Compacting Concrete, A. Skarendahl and O. Petersson, eds., 1999, pp.3-14.
2. Shi, C., Wu, Y. and Riefler, M., "Comparison of Self-Compacting with Conventional Concretes," Proceedings of the 5th International Symposium on Cement and Concrete, Shanghai, China, October 28-31, 2002.
3. Ozawa, K., Maekawa, K. and Okamura, H., "Development of High Performance Concrete," Proceedings of the JCI, Vol. 11, No. 1, 1989.
4. Zhu W., Gibbs J C and Bartos P J M. "Uniformity Of In Situ Properties Of Self-Compacting Concrete In Full-Scale Structural Elements," Cement and Concrete Composites, Vol.23, No. 1, pp.57-64, 2001.
5. Ho, D. W. S., Sheinn, A. M. M. and Tam, C. T., "The Sandwich Concept Of Construction With SCC," Cement and Concrete Research, Vol.31, No. 9, pp. 1377-1381, 2001.
6. Okamura, H. and Ozawa, K., "Mix–Design for Self-Compacting Concrete," Concrete Library of JSCE, No.25, pp.107-120, June 1995.
7. Sedran, T. and de Larrard, F., "Optimization of Self-Compacting Concrete Thanks to Packing Model," Proceedings of the 1st RILEM Symposium on Self-Compacting Concrete, A. Skarendahl and O. Petersson, eds., 1999, pp.321-332.

8. Saak, A. W.; Jennings, H. M. and Shah, S. P., "New methodology for Designing Self-Compacting Concrete," ACI Materials Journal, V.98, No.6, November-December 2002, pp.429-439.
9. Su, N.; Hsu, Hsu, K. C. and Chai, H. W., "A Simple Mix Design Method for Self-Compacting Concrete," Cement and Concrete Research, 31, 2001, pp.1799-1807.
10. Bui, V., Akkaya, Y. and Shah, S. P., Rheological Model for Self-Consolidating Concrete, ACI Materials Journal, Vol.99, No.6, 2002, pp.549-559.
11. Hwang, C. L. and Chen, Y. Y., "The Property of Self-Consolidating Concrete Designed by Densified Mixture Design Algorithm," Proceedings of the 1st North American Conference on the Design and Use of Self-Consolidating Concrete, November 12-13, 2002, pp.121-132.
12. Shi, C., Wu, Y., Shao, Y. and Riefler, M., "Comparison of Two Approaches for Design of Self-Consolidating Concrete", Proceedings of the 1st North American Conference on the Design and Use of Self-Consolidating Concrete (SCC), Northwestern University, Nov 12-13, 2002, pp.349-354.
13. Khayat, K. H. "Viscosity-Enhancing Admixtures for Cement-Based Materials – An Overview," Cement and Concrete Composites, 20, 1998, pp.171-188.
14. Rols, S.; Ambroise, J. and Péra, J., "Effects of different viscosity agents on the properties of self-leveling concrete," Cement and Concrete Research, 29, 1999, pp.261-266.
15. Sari, M.; Prat, E. and Labastire, J.-F., "High Strength Self-Compacting Concrete Original Solutions Associating Organic and Inorganic Admixtures," Cement and Concrete Research, Vol.29, pp. 813-818, 1999.
16. Ferraris, C. F., Obla, K. H. and Hill, R., "The Influence of Mineral Admixtures on the Rheology of Cement Paste and Concrete," Cement and Concrete Research, 31, 2001, pp.245-255.
17. Sakata, K. Ayano, T, and Ogawa, "Mixture Proportioning for Highly-Flowable Concrete Incorporating Limestone Powder," Proceedings of 5th International Conference on Fly Ash, Silica Fume, Slag, and Natural Pozzolans in Concrete, ACI SP-153, 1995, 249-268.
18. Ho, D. W. S., Sheinn, A. M. M., Ng, C. C. and Tam, C. T., "The Use of Quarry Dust for SCC Applications," Cement and Concrete Research, Vol.32, No.4, pp.509-512, 2002.
19. Bouzoubaâ N. and Lachemi M. "Self-Compacting Concrete Incorporating High Volumes of Class Fly Ash; Preliminary Results," Cement and Concrete Research, 31, 2001, pp. 413-420.
20. Xie, Y., Liu, B., Yin, J. and Zhou, S., "Optimum Mix Parameters of High-Strength Self-Compacting Concrete with Ultrapulverized Fly Ash," Cement and Concrete Research, 32, 2002, pp. 477-480.
21. Ghezal, A. and Khayat, K. H., "Optimizing Self-Consolidating Concrete with Limestone Filler by Using Statistical Factorial Design methods," ACI Materials Journal, V.99, No.3, May-June 2002, pp.264-272.
22. Bombled, J. P., "Rheology of Fresh Concrete: Influence of Filter Addition in the Cements," Proceedings, Eighth International Congress on the Chemistry of Cement, Vol. IV, Rio de Janeiro, 1986, p.190-195.

23. Neto, C. S. and Campiteli, V. C., "The Influence of Limestone Additions on the Rheological Properties and Water Retention Value of Portland Cement Slurries," Carbonate Addition to Cement, ASTM STP 1064, P. Klieger and R. D. Hooton, Eds., American Society for Testing and Materials, Philadelphia, 1990, pp.24-29.
24. Lu, P. and Lu, S., "Effect of Calcium Carbonate on the Hydration of C_3S," Guisuanyan Xuebao, Vol. 15, No. 4, 1987, pp 289-294.
25. Peterson, Ö., "Limestone Powder as Filler in Self-Compacting Concrete-Frost Resistance, Compressive Strength and Chloride Diffusivity," Proceedings of the 1st North American Conference on the Design and Use of Self-Consolidating Concrete. 2002, Chicago, pp. 391-396.
26. Adams, L. D. and Race, R. M., "Effect of Limestone Additions Upon Drying Shrinkage of Portland Cement Mortar," Carbonate Addition to Cement, ASTM STP 1064, P. Klieger and R. D. Hooton, Eds., American Society for Testing and Materials, Philadelphia, 1990, pp. 41-50.
27. Raghavan, K. P., Sarma, B. S. and Chattopadhyay, D., "Creep, Shrinkage and Chloride Permeability Properties of Self-Consolidating Concrete," Proceedings, First North American Conference on the Design and Use of Self-Consolidating Concrete. 2002, Chicago, pp. 341-347.
28. Montgomery, D. and Van, B. K., "Drying Shrinkage of Self-compacting Concrete Containing Milled Limestone," Proceedings of the First International RILEM Symposium on Self-Compacting Concrete, Stockholm, Sweden, 1999, pp. 227-238.
29. Soroka, I. and Stern, N., "Calcareous Fillers and the Compressive Strength of Portland Cement," Cement and Concrete Research, Vol. 6, 1976, pp.367-276.
30. Shi, C. and Shao, Y., "What is the Most Efficient Way to Activate Pozzolanic Reactivity of Fly Ashes?" Proceedings of the 30th Canadian Society for Civil Engineering, Montreal, pp. M0-34-1-10, May 2002.
31. Helmuth, R. 1987. "Fly Ash in Cement and Concrete," Portland Cement Association, Skokie, Illinois.
32. Xi, Y. and Jennings, M., "Relationships between Microstructure and Creep and Shrinkage of Cement Pastes," Materials Science of Concrete III, ed by J. Skalny, American Ceramic Society, 1992, pp.37-70.
33. Collins, F. and Sanjayan, J. G., "Effect of Pore Size Distribution on Drying Shrinkage of Alkali-activated Slag Concrete," Cement and Concrete Research, Vol.30, No.9, 2000, pp.1401-1406.

Table 1: Chemical composition of Portland cement and coal fly ash

Oxide	Composition (%)	
	Type I Portland Cement	Class F Coal Fly Ash
SiO_2	20.33	47.8
Al_2O_3	4.65	23.4
Fe_2O_3	3.04	15.1
CaO	61.78	3.36
MgO	3.29	0.81
SO_3	3.63	1.33
Na_2O	0.24	0.72
K_2O	0.59	1.7
Density, kg/m^3	3150	2520

Table: 2 Physical properties of Portland cement and class F fly ash

	ASTM Type I Portland Cement	Fly Ash	ASTM C 618 for Class Fly Ash
Fineness			
passing 325 mesh, %	96.1	73.6	66 (min)
specific surface area (Blaine) (m^2/kg)	383	-	-
Compressive Strength (Mortar Cubes) (MPa)			
1-day	14.3	-	-
3-day	21.8	-	-
7-day	26.1	-	-
28-day		-	-
Strength Activity Index with Portland Cement (%)			
7-day	-	76.7	75 (min)
28-day	-	-	-
Water Requirement (% of cement control)		97.1	105 (max)
Soundness, autoclave expansion or contraction (%)	0.198	-0.02	0.8 (max)

Table 3: Mixture proportions of SCCs

Mix No.	Content (kg/m^3)						
	Cement	Fly Ash	Stone Dust	Fine Aggregate	Coarse Aggregate	Water	SP
SCC-SD	330	-	200	870	750	212	3.7
SCC-FA	330	200	-	870	750	212	3.8

Table 4: Properties of the two fresh SCCs

Mix No.	Slump Flow (mm)	L-Box H2/H1 (%)	L-Box Flow* (s)	Segregation Resistance	Bleeding (visual)	Air Content (%)	Density (kg/m³)
SCC-SD	570	76	3.81	Excellent	No	2.9	2300
SCC-FA	580	79	3.10	Excellent	No	2.2	2300

* flow time from the door to the end of the L-box (800 mm)

Table 5: Setting times of SCCs

Mix No.	Initial Setting Time (h:m)	Final Setting Time (h:m)
SCC-SD	5:00	6:50
SCC-FA	6:15	8:10

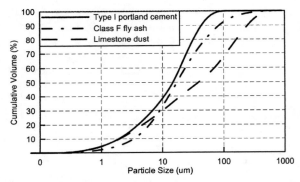

Figure 1: Particle size distribution of portland cement, coal fly ash and crushed stone dust

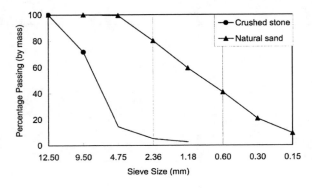

Figure 2: Gradation of aggregates

Recycling Concrete and Other Materials 127

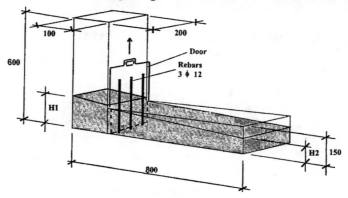

Figure 3: Illustration of L-box

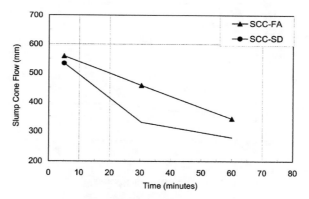

Figure 4: Change of slump cone flowability of SCCs with time

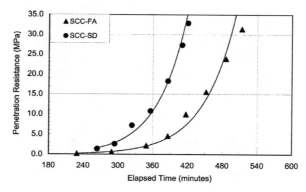

Figure 5: Development of penetration resistance of SCCs

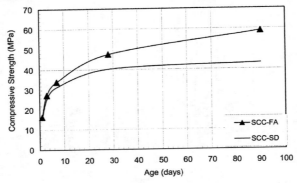

Figure 6: Strength development of SCCs

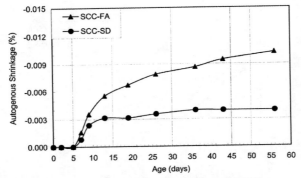

Figure 7: Autogenous shrinkage of SCCs

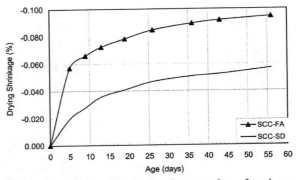

Figure 8: Drying shrinkage of SCCs after one day of moisture curing

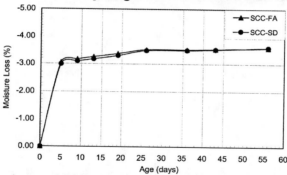

Figure 9: Moisture losses of SCCs during drying testing after one day of moisture curing

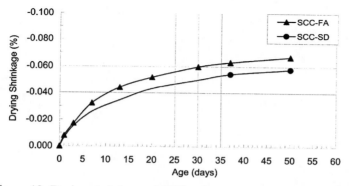

Figure 10: Drying shrinkage of SCCs after seven days of moisture curing

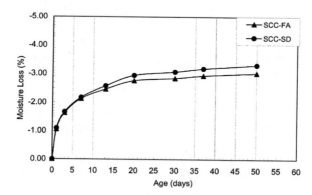

Figure 11: Moisture losses of SCCs during drying testing after seven days of moisture curing

SP-219—10

Recycled Waste Latex Paint as an Admixture in Concrete Sidewalks

by M. Nehdi

Synopsis: Waste latex paint constitutes 12% of the total hazardous waste collected in Ontario, Canada. Currently only 10 to 30% of this waste is being collected. With increasingly more stringent environmental regulations on volatile organic compounds (VOCs), more latex-based paints will be produced compared to solvent- and oil-based alkyds. This will result in more waste latex paint being generated annually in Ontario and across North America. It costs municipalities between $0.90 and $1.40 CAD per litre to dispose of such waste. This study aims at investigating the benefits of recycling waste latex paint in concrete with a special focus on concrete sidewalks. Waste latex paint was used in concrete mixtures both as a partial replacement for virgin latex and for mixing water. It is shown that concrete mixtures incorporating waste latex paint have improved workability, higher flexural strength, lower chloride ion penetrability, better resistance to deicing salt surface scaling and could be more economic because they require less water-reducing and air-entraining admixtures. The annual urban concrete sidewalk construction could use the yearly production of waste latex paint while producing sidewalks with enhanced properties and durability.

Keywords: chloride ions; concrete; latex paint; recycling; sidewalks; surface scaling; waste

Nehdi

Moncef Nehdi is Associate Professor in the Department of Civil and Environmental Engineering, University of Western Ontario. He is a member of ACI committees 225 Hydraulic Cements, 236 Material Science of Concrete, 555 Recycling, and 803 Faculty Network Coordinating. His research interests include modeling the behavior of cement-based materials, recycling by-products in construction, and durability and repair of concrete infrastructure. Nehdi is recipient of the 2003 ACI Young Member Award for Professional Achievement.

INTRODUCTION

To deal with an escalating waste disposal crisis, the Ministry of Environment in Ontario implemented the 3R's management strategy (reduction, reuse, and recycling). Consequently, the collection of hazardous waste (HZW), including waste paint, has become a common practice for several years. For example in 1999, 6,364 t of HZW was collected in the province, among which 21.7% is waste paint. Waste latex paint (WLP) alone constitutes 12% (around 500,000 L) of the total HZW and has been accumulating since inception of its collection policy. It is believed that the collected WLP will grow substantially because current collection programs result in the capture of only 10 to 30% of the total HZW. Public education programs can raise this level significantly. Moreover, increasingly more stringent environmental regulations on volatile organic compounds (VOCs) will increase the production and use of latex paints relative to other solvent-based paints, resulting in an increase of WLP.

Laboratory tests on recycling WLP in asphalt concrete provided positive results. However, field pavement trials indicated that unpleasant odours were emitted from the mixture. Air quality analyses showed that hot mix asphalt mixtures incorporating recycled latex paint do release vapours of ethylene glycol in concentrations beyond the recommended levels for occupational health and safety. There was no practical and economical way of stripping the harmful vapours out of asphalt concrete. Therefore, this option was terminated (1).

This study aims at investigating the use of WLP as an admixture in portland cement concrete. The idea of using polymers in cement-based materials dates back to the early 1920s when the first patents on using natural rubber polymer-modified cementitious systems were issued (2). The first patent on the use of synthetic rubber latexes in such application was issued in 1932 (2). Since then many products, patents and applications have been developed (2). In North America, latex-modified concrete (LMC) was used as a bridge overlay in Michigan as early as in 1958. In Ontario, the first major application of LMC was a 1980 overlay on collector lanes of Highway 401 in North York. However, information on the use of WLP in concrete was not accessible in open literature, and it is not understood whether WLP could contribute improvements to the behaviour of concrete similar to those imparted by virgin latexes.

WLP can be used as a value added product for partial replacement of virgin latex in LMC. This requires that the beneficial effects of using this by-product must be equivalent to those of virgin latexes. However, LMC is used in demanding applications such as bridge deck overlays to mitigate corrosion problems. This implies that stringent quality control procedures must be implemented to assure that the performance of a variable and non-controlled by-product such as WLP match that of a factory-controlled product, which is difficult to achieve. A simpler approach is to use the WLP as partial replacement for mixing water in concrete that is normally not intended to be polymer modified. The use of the recycled WLP must improve or maintain equal performance of the host concrete. Concrete sidewalks seem to be a good application for the latter case since the reduction of permeability and porosity due to the latex polymer may enhance the resistance of concrete to surface scaling and other physical and chemical attack. The yearly WLP collected in Ontario can be recycled in about 9,000 m^3 of concrete, which can roughly make 48 km of sidewalk (assuming 50 L of WLP per m^3 of concrete, each m^3 of concrete can make 5.3 m of sidewalk 1.5 m wide and 0.125 m deep). Therefore, the concrete sidewalk industry can potentially recycle most WLP each year. In the present study, WLP was used both as partial replacement for virgin latex and as partial replacement for mixing water. Fresh concrete properties, compressive strength, flexural strength, rapid chloride penetrability and resistance to deicing salt surface scaling were evaluated for various concrete mixtures and results are discussed in this paper.

MATERIALS AND TEST METHODS

CSA Type 10 (ASTM Type I) cement, slag, coarse and fine aggregates, air-entraining admixture and water-reducing admixture were obtained from typical materials currently used in municipal sidewalk concrete mixtures in London, Ontario. Various samples of WLP from the City of London's waste collection site were obtained and analysed. Their properties are summarized in Table 1. Generally, the latex polymers represent about 15% of WLP, water is around 59%, and other extender solid pigments including TiO_2 are around 25%. These proportions vary significantly in time but considering that the material is a non-controlled by-product, the variation shown in Table 1 is considered acceptable. However, the material is expected to have more uniform properties if larger container mixing is implemented to average out any variability. The virgin latex used was a styrene-butadiene with a specific gravity of 1025 kg/m^3, a non-volatile content of 47%, which generally corresponds to the solid latex polymer content (note that this is significantly higher than the polymer content in the latex paint).

To investigate the partial replacement of virgin latex by recycled latex paint, a reference concrete mixture with a water-cement ratio (w/c) of 0.40 and a latex modified mixture with similar w/c were made. In addition, mixtures containing 100, 75, and 25% replacement of the virgin latex by recycled latex paint were prepared, and their proportions are shown in Table 2. Although the polymer/cement ratio will not be maintained when recycled latex paint is used as partial replacement for virgin latex on a mass basis, this option was adopted to make the replacement procedure simpler. The mixing water was adjusted to account for the additional water in the latex and recycled paint. Moreover, to investigate the effect of using WLP in municipal sidewalk concrete mixtures that are normally not intended to be modified with polymer, a reference sidewalk concrete mixture and mixtures having 10, 20, 30, 40 and 60% replacement of mixing water with recycled latex paint were made and their proportions are shown in Table 3. Water and WLP and/or virgin latex were manually mixed. The saturated surface-dry coarse aggregates were mixed with part of the mixing water (with WLP and/or latex). The fine aggregates and cementitious materials were subsequently added and mixing resumed for 2 min with the remaining mixing water and chemical admixtures being added over the first minute of mixing. After a 2-min rest period, a 3-min mixing sequence was conducted, and the mixer was covered with a plastic sheet.

The slump (ASTM C 143) and air content (ASTM C 231) of fresh concrete were measured. Cylindrical specimens 100x200 mm, beams 100x100x350 mm and slabs 200x230x75 mm were prepared using a vibrating table. Specimens without latex were covered by wet burlap for 24 h, and then cured in a moist curing room at 23°C and more than 95% RH. Specimens containing latex and/or WLP were covered with wet burlap for 48 h then air-cured under lab conditions at around 23°C until testing. The different regime of curing for the specimens incorporating latex is needed to allow the polymerization of the latex monomers in such specimens (2). For each mixture, the compressive strength was measured on three cylindrical specimens at 1 d and 28 d according to ASTM C 39 specifications. The flexural strength was obtained on three beam specimens as per ASTM C 78. In addition, the rapid chloride penetrability was obtained at 28 d on three disks 50x100 mm cut from cylindrical specimens 100x200 mm for each mixture as per the ASTM C 1202 guidelines. After curing the slabs (14 d in moist chamber and 14 d in lab conditions for slabs with no latex; 1 d moist curing and 27 d in lab conditions for slabs with latex), strips of high density expanded polystyrene were attached to the sides of slabs with an exterior grade adhesive in order to create a watertight brine pond with a 6 mm thickness and a concentration of 4% $CaCl_2$. Starting at an age of 28 d, the slabs were subjected to 50 cycles of freezing-thawing strictly following the ASTM C 672 guidelines. The amount of scaled-off material was measured after each 5 cycles, and scaling was given a visual rating. The results are average values obtained on two identical slabs for each mixture.

RESULTS AND DISCUSSION

The air content and slump results for mixtures using WLP as partial replacement for virgin latex and mixing water are shown in Tables 2 and 3, respectively. It is observed that the air content of fresh concrete increased with the level of replacement of mixing water with WLP. This behaviour is usually observed with virgin latexes, and is believed to be due to surfactants contained as emulsifiers and stabilizers in polymer latexes (2). Commercial latexes intended for use in concrete often contain antifoaming agents to reduce air entrainment. It was decided to reduce the dosage of the air-entraining admixture and add an antifoaming agent in the recycled paint (mixtures with 30-60% WLP in Table 3), which decreased the air content, but this effect was unpredictable and difficult to control. When the latex paint was used for partial replacement of virgin latex, it was coupled with an antifoaming agent and the air entrainment was better controlled.

It is also observed that as the paint replacement level increased, the slump of the mixtures increased (Table 3), especially when coupled with a high air content. This effect allowed eliminating the water reducer in sidewalk concrete mixtures, which could yield significant cost savings. It could also allow reducing the w/c, thereby enhancing strength and durability of concrete. Improvements in workability due to WLP addition can be attributed to the ball-bearing effect of polymeric materials (2), the increased entrained air content, the dispersing effect of surfactants present in WLP, and the effect of ultrafine pigments used as extenders.

One difficulty associated with using WLP is that the surfactants used in its manufacturing are generally different from those used in making latex supplied for latex modified concrete. Thus, predicting its effect on air entrainment properties and workability is difficult, and needs careful assessment. For industrial-scale implementation, it is recommended that WLP be collected in large containers that are equipped with a pump circulating system. The continuous mixing action will reduce the variability of the material, and prevent settlement of the solid phase of the paint, which may otherwise introduce unacceptable variability in concrete mixtures. Quality control tests on WLP could be regularly conducted which allows altering concrete mixtures accordingly.

Table 4 shows compressive strength, flexural strength and rapid chloride penetrability test results for concrete mixtures incorporating various proportions of WLP as partial replacement for virgin latex. The compressive strength at 1 d and 28 d increased by around 15% and 20%, respectively compared to those of a reference mixture when virgin latex was used at a polymer/cement ratio of 15%. As the proportion of WLP replacement of virgin latex increased from 0 to 100%, the compressive strengths decreased both at 1 d and at 28 d (Table 4), but generally remained higher or comparable to that of the reference mixture except the 1 d strength results of mixtures with 75 to 100% of WLP which were lower likely due to a set retarding effect of the WLP. A significant increase in compressive strength of concrete due to polymer modification is usually not noticeable (2). However, some mixtures (Table 4) showed higher 28 d strength than that of the reference mixture probably due to combined effects of improvement in workability and the filler effect and/or pozzolanic activity of the pigments used in WLP. A decrease in the strength of some mixtures containing WLP may be attributed to an increase in the air content measured at high WLP dosages, though the mixture with the highest air content in Table 4 still achieved the second highest strength. It could also be that the concretes with high WLP content are more sensitive to air curing, but this needs further investigation.

Table 4 also shows that the flexural strengths at 1 d and 28 d increased by around 27% and 17%, respectively compared to those of a reference mixture when virgin latex was used at a polymer/cement ratio of 15%. These improvements are usually due to the polymer itself and to an overall enhancement of the aggregate–cement paste bond (2). The flexural strength of mixtures incorporating WLP was lower than that of the reference mixture at 1-d, but exceeded the 28 d flexural strength of the reference mixture when set-retarding effects were overcome. It can be noted that when the ratio of WLP/virgin latex paint increased, the enhancement in flexural strength decreased. This is likely due to the fact that the polymer content in the virgin latex is about three times higher than that in WLP. The amount of polymer available in the concrete mixtures decreased with higher WLP proportion and so did the beneficial effect on flexural strength.

Figures 1 and 2 show the 28 d rapid chloride penetrability test results for mixtures incorporating virgin latex and/or WLP relative to that of a reference mixture along with the corresponding rating ranges of ASTM C 1202. Using virgin latex in concrete at a polymer/cement ratio of 15% decreased the rapid chloride penetrability of concrete by around 55% from a moderate to a low rating. Although replacing virgin latex with waste latex paint would decrease the polymer/cement ratio (WLP contains only 1/3 of polymer compared to virgin latex), it decreased the rapid chloride penetrability by 45%, a value comparable to that achieved by virgin latex. Furthermore, a 75% virgin latex-25% WLP was optimal, and reduced the rapid chloride penetrability of concrete by 65%. The reduction of chloride ion penetrability due to virgin latex is due to the pore blocking effect of polymers. However, WLP contains only one third of the polymers in virgin latex. It is believed that the extender pigments used in latex paints played a significant role in this regard. These extender pigments are often present at around 25% by mass, and contain a blend of titanium dioxide, calcium carbonate, calcinated clays, diatomaceous earth, metakaolin, etc. (3). Some of these mineral pigments significantly decrease chloride ion penetrability through mechanisms similar to those of silica fume.

Table 5 shows compressive strength, flexural strength and rapid chloride penetrability test results for sidewalk concrete mixtures incorporating various proportions of WLP used as partial replacement for mixing water. The 1 d compressive strength decreased with higher WLP dosage due to increased air content and a set-retarding effect. It is known that the setting of polymer-modified concrete is somewhat delayed compared to normal concrete (2), and this effect is dependent on the polymer type and polymer-cement ratio. However, the 28 d compressive strength of mixtures having air contents comparable to or lower than that of the reference mixture reached or exceeded the strength of the reference mixture mainly due to lower w/c. Also, such mixtures achieved higher 28 d flexural strength results than the reference mixture. As shown in Fig. 2, WLP achieved substantial decreases in chloride ion penetrability of concrete mixtures, and this effect was higher the higher the dosage. At 60% replacement of mixing water by WLP, the chloride ion penetrability decreased by 65% from a high rating to a low rating (ASTM C 1202). It is worth mentioning that as the replacement level of mixing water by WLP on a mass basis increases, the effective w/c ratio decreases (WLP contains 59% water), which helps decreasing chloride penetrability, in addition to the polymer and extender pigment effect mentioned earlier. It is worth mentioning in this context that the WLP used as partial replacement for virgin latex and that used as partial replacement for mixing water were sampled at different dates. Thus, their effects on workability, air content, strength and other properties cannot be directly compared.

The results of deicing salt surface scaling tests are summarized in Table 6. Using virgin latex at a polymer/cement ratio of 15% enhanced the surface scaling rating of concrete slabs from 2 (no latex) to 1 (with latex). Replacing 40% of the mixing water by waste latex paint achieved comparable results. This could be beneficial for instance when WLP is used in municipal sidewalk concrete mixtures. Figure 3 shows the deicing salt surface scaling behaviour after 50 freezing-thawing cycles of a reference concrete slab and a slab incorporating 40% replacement of mixing water with WLP. It is shown that unlike the slab made from the reference mixture, the one made from the WLP-modified mixture did not exhibit aggregate exposure. It is noted that the curing regime for the latex modified mixtures (27 d in lab-air) was different from that of the reference mixture (cured at RH>95%), and this could be an additional factor in the observed surface scaling performance.

The above results indicate that positive effects can be expected when WLP is used for partial replacement of mixing water in sidewalk concrete mixtures. An experimental section of a municipal sidewalk was constructed in 1998 at the City of London's waste collection site, and its performance is being monitored (Fig. 4). The mixture contained 375 kg/m^3 of CSA Type 10 cement, 715 kg/m^3 of natural sand, 1080 kg/m^3 of limestone coarse aggregate and 50% replacement of mixing water with recycled WLP. Its air content was 6%, and its 28 d compressive strength reached 35 MPa. The experimental section included a 5.0x1.7 m concrete sidewalk with a regular finish, a 2.0x1.7 m sidewalk with an impressed concrete finish and a 15-m long concrete curb and gutter. In addition to the ease of construction observed when WLP was added to concrete, the sidewalk had a distinctly lighter and more reflective color compared to a reference sidewalk, which could be an advantage for night vision. It also did not exhibit coarse aggregate pop-outs present in the reference sidewalk, though this problem is usually more related to the aggregates than to the surrounding cement paste. Although deicing salts have been used on the experimental sidewalk, no signs of surface scaling have been observed to date. The demonstration WLP-modified sidewalk had much higher quality finish, and the impressed concrete sidewalk showed much sharper details at the edges of patterns, indicating substantial gains in the finishing process.

Various issues need more research before recommending a large-scale industrial use of WLP in concrete. The conventional generic polymers used in virgin latexes supplied for LMC are often acrylic or styrene-butadiene rubber, which are known to reduce permeability and resist alkaline breakdown. The mixture of polymers found in WLP is predominantly copolymers of acrylics and vinyl acetate. The long-term performance of these polymers in alkaline environments needs further investigation. In addition, the effect of surfactants and antifoaming agents present in WLP on the spacing factor of air bubbles in air-entrained concrete can be critical for its freezing-thawing and surface scaling durability, and needs laboratory and field examination. No measurements of the alkali content in recycled paints have been conducted, and their effect on expansion of concrete due to alkali silica reaction needs clarification. Technological issues related to quality control, limiting effects of contaminants and variability in time of WLP need further investigation. In addition, the concrete industry may need special equipment or special maintenance procedures to handle concrete incorporating large amounts of recycled paint. Leaching tests are also needed to investigate whether solvents present in WLP are fixed in concrete or can migrate to the outside environment. Such issues need further research.

CONCLUSIONS

The use of waste latex paint as partial replacement for virgin latex in latex-modified concrete and as partial replacement for mixing water in municipal sidewalk concrete mixtures was examined in this study. It was found that WLP contributes a significant part of the advantages of virgin latexes in concrete, such as increasing flexural strength and decreasing chloride ion penetrability. A field demonstration sidewalk modified with WLP exhibited better workability and finishing, more appealing color, and better resistance to surface scaling and aggregate pop-outs than a reference sidewalk. More research is needed to investigate the effect of WLP on the stability and spacing factor of air-entrained concrete, the effect of WLP on expansion due to alkali silica reaction, the variability of WLP in time, its effect on industrial concreting equipment and the stability of contaminants that may be present in the recycled paint.

REFERENCES

(1) Earth-Tec Canada Inc., City of London – Environmental Services, CBM, Ashwarren Engineering and University of Western Ontario, "Waste Latex Paint Re-Use Project", Final report, prepared for The Waste Diversion Organization, and City of London, June 22, 2001.

(2) Ohama, Y., "Polymer-Modified Mortars and Concretes", in Concrete Admixtures Handbook: Properties, Science and Technology, Edited by V.S. Ramachandran, Noyes Publictions, 1984, pp. 337-422.

(3) Wicks, Z.W. Jr., Jones, F.N. and Pappas, S.P., Organic Coatings, Vol. I, John Wiley & Sons, 1994, 343 p.

Table 1 - Typical compositions of waste latex paint

Date sampled	Specific gravity (kg/m^3)	Non volatile content (%)	Ethylene glycol (%)	Water (%)	Latex polymers (%)	TiO$_2$ (%)	Extender pigments (%)
12/02/00	1.53	47.4	1.0	52.6	16.5	15.5	15.5
12/02/00	1.43	37.6	1.0	62.4	14.2	11.7	11.7
01/27/01	1.51	42.5	1.2	57.5	12.9	14.8	14.8
01/27/01	1.46	40.6	1.4	59.4	14.7	12.9	12.9
02/03/01	1.47	44.5	1.4	55.5	18.1	13.2	13.2
02/24/01	1.46	40.2	1.0	59.8	14.2	13.0	13.0
02/24/01	1.41	38.2	1.1	61.8	16.3	10.9	10.9
03/03/01	1.38	35.8	0.8	64.2	17.1	9.4	9.4
03/03/01	1.41	39.4	0.9	60.6	17.5	10.9	10.9
Average	1.45	40.7	1.1	59.3	15.7	12.5	12.5
Std. Dev.	0.05	3.4	0.2	3.4	1.7	1.8	1.8

Table 2 – Mixture proportions for replacing virgin latex with waste latex paint and properties of fresh concrete

	Ref. mixture	100% Virgin latex	75% latex-25% paint	25% latex-75% paint	100% paint
Cement (kg/m^3)	390	390	390	390	390
Water (kg/m^3)	155	90	90	85	80
Coarse aggregate (kg/m^3)	890	890	890	890	890
Fine aggregate (kg/m^3)	820	820	820	820	820
Virgin latex (kg/m^3)	--	125	95	30	--
Recycled paint (kg/m^3)	--	--	30	95	125
Slump (mm)	75	105	110	115	120
Air content (%)	6.5	8.5	9.5	8.0	9.0
Polymer-cement ratio (%)	-	15.0	12.7	7.4	5.0
Water-cement ratio (%)	0.40	0.40	0.41	0.40	0.39

Table 3 – Mixture proportions for replacing mixing water with waste latex paint and properties of fresh concrete

	Ref. mix	10% paint	20% paint	30% paint	40% paint	60% paint
Cement (kg/m^3)	285	285	285	285	285	285
Slag (kg/m^3)	70	70	70	70	70	70
Cementitious materials (cement + slag) (kg/m^3)	355	355	355	355	355	355
Coarse aggregate (kg/m^3)	1070	1070	1070	1070	1070	1070
Fine aggregate (kg/m^3)	750	750	750	750	750	750
Water (kg/m^3)	135	125	110	95	82	55
Waste latex paint (kg/m^3)	--	14	27	40	55	82
Water-reducing admixture (mL/100 kg of cement)	220	220	220	220	0	0
Air-entraining admixture (mL/100 kg of cement)	10	10	10	5	5	5
Slump (mm)	58	160	180	180	120	110
Air content (%)	8.5	10.5	16	12	7	4
Polymer-cement ratio (%)	-	0.6	1.2	1.8	2.4	3.6
Water-cementitious materials ratio (%)*	0.38	0.38	0.35	0.33	0.32	0.29

* mass ratio of (mixing water + water from waste latex paint)/cementitious materials. This ratio decreased with higher WLP content because replacement of mixing water with WLP was based on mass and only about 59% of WLP is water.

Table 4 – Experimental results: replacement of virgin latex with WLP

Mixture	1 d f'$_c$ (MPa)	28d f'$_c$ (MPa)	1 d flexural strength (MPa)	28d flexural strength (MPa)	28d Chloride penetrability (C)
Reference	13.7	44.5	6.1	8.6	1880
100% Virgin latex	17.2	53.3	7.8	10.2	850
75%latex-25% WLP	16.4	51.6	5.4	10.5	470
25%Latex-75% WLP	11.2	48.9	5.6	9.8	1350
100 WLP	8.7	44.7	5.8	9.2	1040

Table 5 – Experimental results: replacement of mixing water with WLP

Mixture	1 d-f'_c (MPa)	28d-f'_c (MPa)	28d flexural strength (MPa)	28d Chloride penetrability (C)
Reference	15.4	35.8	5.8	2130
10% WLP	11.4	27.6	5.6	1750
20% WLP	6.6	17.0	5.0	1900
30% WLP	9.5	22.0	5.3	1250
40% WLP	8.6	41.1	6.1	990
60% WLP	7.5	44.3	6.4	750

Table 6 – Results of deicing salt surface scaling tests

Mixture	Visual rating ASTM C672	Total scaling residue (kg/m^2)
Control (no latex), w/c ratio = 0.40	2	0.85
Virgin latex polymer-cement ratio = 15%, w/c ratio = 0.40	1	0.55
40% replacement of mixing water with waste latex paint polymer-cementitious material ratio = 2.4 %, w/b ratio = 0.36	1	0.60
25% virgin latex – 75% waste latex paint polymer-cement ratio = 7.4%, w/c = 0.40	1	0.45

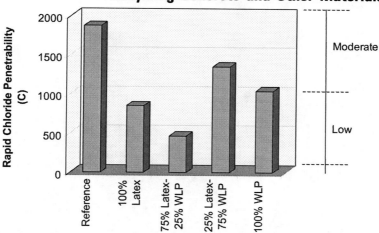

Fig. 1 - Rapid chloride penetrability of concrete mixtures incorporating WLP as partial replacement for virgin latex.

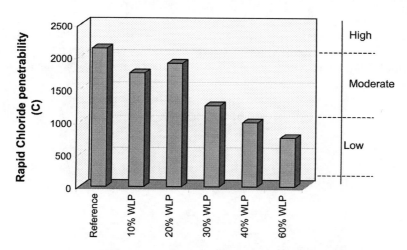

Fig. 2 - Rapid chloride penetrability in sidewalk concrete mixtures incorporating various proportions of waste latex paint as partial replacement for mixing water.

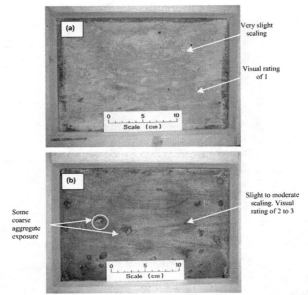

Fig. 3 - Deicing salt surface of (a) reference concrete slab (moderate scaling, visual rating of 2 to 3) and (b) slab from a concrete incorporating 40% replacement of mixing water with WLP (slight scaling, visual rating of 1) after 50 freezing-thawing cycles.

Fig. 4 - Section of an experimental concrete sidewalk incorporating 50% waste latex paint as partial replacement for mixing water.

Development of a New Binder Using Thermally-Treated Spent Pot Liners from Aluminium Smelters

by A. Tagnit-Hamou and S. Laldji

Synopsis: The use of mineral admixtures as addition to or replacement of portland cement has been attracting a great amount of interest in recent years. Using suitable quantities of those minerals not only improves some properties of fresh and dry concrete, but also reduces portland cement demand and helps solve several environmental problems. Aluminium production in various parts of the world generates a considerable amount of waste which contains leachable cyanides and fluorides that cause a serious environmental problem.

This paper presents a study of the effect of substituting a percentage of cement by a glass frit in mortar and in concrete. The term glass frit refers to spent pot liners resulting from the aluminium production process, that have undergone various treatments and have been ground to the fineness of a cement. The various results obtained in different tests conducted on mortar and concrete showed that glass frit has a remarkable reactivity potential and an interesting rheological behaviour. Replacing a percentage of cement by glass frit improves workability and strengths of mortar and concrete. For a given slump, concrete containing 25% of glass frit requires 50% less water reducer than that of control concrete. The compressive strengths developed in mortar or in concrete are very similar or even greater than those made with portand cement only or those incorporating blast furnace slag with equivalent cement replacement rate.

Keywords: concrete; glass frit; mineral admixture; spent pot liner; sustainable development

Tagnit-Hamou and Laldji

ACI member Arezki Tagnit-Hamou is a Professor in the Department of Civil Engineering at the Université de Sherbrooke, Sherbrooke, Québec, Canada. His research interests include physico-chemistry, microstructure and durability, supplementary cementitious materials, and sustainable development.

ACI member Said Laldji is a research assistant in the Department of Civil Engineering at the Université de Sherbrooke, Sherbrooke, Québec, Canada. His research interests include supplementary cementitious materials, sustainable development, and expansion due to alkali silica reaction.

INTRODUCTION

Since the Roman Empire, mineral admixtures have been used in cement. Research carried out in different laboratories have led to a wide variety of blended cements. Concrete containing pozzolanic or cementitious materials exhibit superior durability and better strength perfomance [1,2,3,4]. In addition, and because of environnemental and economic concerns, the worlwide trend today is toward utilization of pozzolanic and mineral admixtures, either in the form of blended cements or as direct additions to portland cement concrete during mixing operations.

The material used in this investigation has never been used in any other work. The main objective of this preliminary study is to improve the value of recycled spent pot liners (SPL). A recent survey carried out in 2000 [5], has shown that the world annual production of primary aluminium has reached more than 8 million tons. Depending on the fabrication technology in use in different industries, 100 tons of aluminium can generate up to 2.5 tons of SPL. As with many other industrial by-products, spent pot liners generated by aluminium plants are not always recycled and re-used. After thermal treatment, cyanides are completely removed from the material. The fluorides are fixed by vitrification in the quenched glassy material which is identified as glass frit (GF).

In this paper, a thermally-treated SPL is evaluated for its cementitious or pozzolanic potential by incorporating it in mortar and in concrete.

This work involves the rheological and mechanical characterization of the material. Various percentages of the obtained ground material, at different water to binder ratios, were studied in replacement of cement by mass.

EXPERIMENTAL PROGRAM

Material characterization

Characterization of the material includes its chemical composition, X-ray diffraction (XRD), density, fineness, and laser particle size distribution

Chemical analysis
Except for the fluorine, which was detected by chromatography, most of the other components were determined using the atomic absorbtion test (AA).

X-ray diffraction
The mineral or glass composition was studied by X-ray diffraction (XRD)

Density
The density of the material was obtained using a picnometer known as a Multipycnometer, which measures the displacement of helium gas through a sample with very small pores under pressure.

Fineness
The fineness test was carried out according to ASTM C204. The test, also called Blaine test, consists in measuring the time required for a given volume of air to pass through a given volume of a compacted powder placed in a cell.

Laser particle size distribution
Grading or particle size distribution of the material wass determined using a laser analyzer type Analysette 22 produced by the Fritsh Company.

Paste tests

Mixes were prepared to evaluate the fluidity of a paste containing 0, 10, 25 and 45 % of GF with a water-binder ratio (w/b) of 0.45, using the mini-slump test (6). Paste-extracted solution was also analyzed by inductively coupled plasma-mass spectrometer (ICP-MS) and the main elements that compose the solution were detected.

The temperature of hydration of a paste containing 0, 25 % GF and 25 % blast furnace slag (GGBFS) as cement replacement with a w/b = 0.45 was measured by adiabatic calorimetry.

Mortar tests

Two series of mixes were prepared to examine the activity index of the glass frit. Graded Standard Sand (Ottawa sand) meeting the requirements of ASTM C778 and CSA type 10 cement were used to prepare the mortars. The mixing process of the mortar was carried out according to ASTM C109.
The standarized activity index test (ASTM C311) was prepared using 20 % replacement of cement by glass frit and compared to that of blast furnace slag of the same fineness with equivalent cement replacement. A water to binder ratio of 0.485 was used for the control mortar and a given slump was obtained through the mini-slump test. This ratio was then adjusted in the other mixes in order to obtain the equivalent slump.

The second series of tests consisted in using a constant water to binder ratio of 0.45 to investigate the effect of the rate of cement replacement by glass frit on compressive strength. Four percentages 0, 10, 25 and 45 % of glass frit and blast furnace slag were incorporated in the mixes in replacement of cement.

Tests on alkali-silica reaction according to ASTM C1260 were conducted in order to detect the potential for deleterious alkali-silica reaction of aggregate in mortars bars. A control mix with 100 % cement as well as a mortar containing 25 % cement replacement by glass frit were prepared. Three mortar bars were moulded for each batch. The storage procedure was carried out according to the specifications quoted above. After the zero reading, each series of mortar bars was totally immersed in 1N NaOH solution and stored at 80°C.

Concrete tests

Three concrete mixes were studied. The mix designs are presented in Table 1. The first mix was used as the control concrete with 100 % portland cement, the second was made with 25 % replacement of cement by glass frit, and the third was made with 25 % replacement of cement by blast furnace slag. CSA Type 10 cement (equivalent to ASTM type I), alluvial sand, and 14 mm and 20 mm aggregate provided by a local company, were used to prepare the concrete. An air-entraining agent and a water reducer were also added.

The alkali-silica reaction test was carried out according to CSA 23.2-14A. Four concrete mixes were prepared using alluvial sand and reactive Spratt aggreagate. Two of theses mixes were made with no additionnal NaOH in the mixing water, whereas the other two were prepared with an increased Na_2O_{equi} content of up to 1.25%, meeting the specification by considering the glass frit as a portland cement having the same alkali content. The concrete prisms were placed in storage containers in which an absorbent material was placed around the inside wall with air tight lids at 38°C, according to the CSA specification (CSA 23.2-14A).

RESULTS AND DISCUSSION

Characterization of the glass frit powder

GF, GGBFS, and portland cement chemical compositions are presented in Table 2. Most of the main elements (SiO_2+ Al_2O_3+ FeO_3 + CaO) are found in the glass frit. Compared to the chemical composition of GGBFS, GF has a low CaO content. The C/S content of GF is very low (0.46) and according to the published litterature, the increase in hydraulicity of slag is closely retated to its CaO and Al_2O_3 contents (4). Smolczyk (7) found that acid slags with C/S < 1, which have previously been considered poorly reactive slags, showed good properties when the low CaO content had been compensated by an increase in MgO and Al_2O_3 content. A similar conclusion can be made for GF. Additionally, the high alkali content of GF may contribute to improving its reactivity but could also affect the alkali-silica reaction, which is currently carefully examined.

The GF was ground in our laboratory facilities with a ball mill. The Blaine fineness and density of the glass frit are 410 m^2/kg and 2.85, respectively. Figure 1 shows that the glass frit has approximately the same particle size distribution as the blast furnace slag with a fineness of 400 m^2/kg and a density of 3.00.

X-ray diffraction analysis shows that the spectrum of GF fits that of blast furnace slag very well (Figure 2). They are both amorphous materials with a halo at 31° 2θ. The spectrum of glass frit shows no evidence of crystalline phases compared to the slag used.

Paste

Mini-slump test

Results of the mini-slump test conducted after 60 minutes of hydration are presented in Figure 3. After 60 minutes, the slumps developed by the paste incorporating 10, 25 and 45 % GF as cement replacement, are 16, 38 and 55 % higher than that of the control paste, respectively.

Adiabatic calorimetry

Figure 4 shows the evolution of the heat of hydration of the paste over time with a water binder ratio of 0.45, as measured in adiabatic cells. As expected, the paste containing 25 % glass frit released less heat compared to the control paste and was very close to that of 25 % blast furnace slag.

Mortar

In the activity index test, the mean compressive strengths obtained on 3 cubes made of mortars of the same slump (ASTM C311) containing 0 and 20 % of cement replacement by glass frit were measured (Table 3). According to ASTM C618, the required strength activity index at 7 and 28 days should meet 75 % of the control.

The strength activity index of mortar containing 20 % GF were 92 % at 7 days, 99 % at 28 days and 107 % at 100 days (Table 4). Those of mortar with 20 % GGBFS were 83 %, 92 %, and 94 %, respectively. Evidently, GF seems to perform better than GGBFS.

This performance of GF compared to slag was mainly due to the water demand required to prepare the mortars. For the same slump, the water to binder ratio in mortar with glass frit is 0.46 whereas that of the control and the mortar containing slag was 0.485 and 0.47, respectively.

With a constant water to binder ratio of 0.45 and at early age, discarding the percentage of cement replacement, the mortar with GF presented similar strengths to those of mortar with slag but slightly lower than those of the control mortar (Figure 5). After 28 days of curing, the 10 % slag mortar developed higher strengths, which exceeded even those of the control mortar. However, with 25 and 45 % replacement, GF shows better performance compared to that of the control. The compressive strengths in 25 % GF mortar increased from 13.6 MPa at 1 day to 44.9 MPa at 28 days; this represents a gain of 31.3 MPa. This is comparable to the gain in GGBFS mortar (32.2 MPa). In mortar with 45 % cement replacement, GF shows an even better gain. The strengths increased from 8.3 MPa at 1 day to 41.1 MPa at 28 days, which gives a gain of 32.8 MPa greater than that of slag mortar with only 28.9 MPa. To better illustrate this behaviour, the ratio of glass frit mortar strength to slag mortar strength at different ages is given in Table 5.

It appears clearly that with 25 and 45 % cement replacement, mortars containing GF developed approximately similar strengths as those of blast furnace slag. Knowing that GF can considerably improve the rheology of the mortar, it follows therefore that for an identical slump the mortar containing GF would require less water which thus can increase strengths even more.

Fresh concrete

Results on fresh concrete, as well as the amount of chemical admixtures needed to adjust the slump and air content, are given in Tables 6. As expected, for a given slump and at a constant W/B ratio (0.45), the concrete containing glass frit required a lower dosage of water reducer (about half) and approximatively the same amount of air-entraining agent as that of the control concrete and the concrete with slag.

The initial and final setting times obtained according to ASTM C403 are presented in Table 7. The setting time of the concrete containing GF was pratically similar to that of the control concrete and slightly smaller than that of the concrete with slag. However, the GGBFS concrete showed a slight retarding period of about 50 minutes, which is mainly due to the higher water reducer dosage. Generally, it is well known that water reducing agent, depending on the dosage added during concrete mixing, can affect significantly the setting time. As pointed out earlier, control concrete and concrete containing slag consume twice as much water reducer as the concrete with glass frit.

Hardened concrete

The mean compressive strength measured on three cylinders (100 × 200 mm) at different ages of curing in moist room are given in Figure 6.
Incorporating glass frit into concrete confirms the previous results obtained on mortar. At 1 day, the control concrete showed greater strength compared to the GF or GGBFS concrete. With a longer period of curing, the concrete with 25 % GF developed approximatively identical strengths as those of control or GGBFS concrete.

The resistance to chloride ion penetration, which provides a rapid indication of the permeability of the concrete, was substantially improved by the incorporation of GF (Figure 7). Concrete containing GF developed a permeability, defined as the charges passed through the sample in coulombs, much lower than that of the control. From 4 to 23 weeks curing in a moist room, the charge decreased from 2600 to 1700 coulombs compared to that of the control concrete that decreased from 5600 and 2500 coulombs. According to ASTM C1202, concrete with a charge ranging from 1000 to 2000 coulombs is considered a low permeability concrete whereas that ranging from 2000 to 4000 indicates a moderate permeability concrete.

Expansion of mortar bars and concrete due to alkali-silica reaction

The mean expansion developed by each mortar is shown in Figure 8. It appears clearly that, after more than 80 days of curing in 1N NaOH solution at 80°C, mortar with 25 % glass frit developed a smaller expansion than that of the control. However, as pointed out in ASTM C1260, at the normalized curing period of 16 days, both mortar containing GF and the control showed expansions exceeding the limit of 0.20% set by the specification. This is indicative of potentially deleterious expansions of the aggregate.

Similar tests carried out by Hudec and Ghamari (8) and Kojima et al. (9) on glass powder having an alkalis content equivalent to that of GF showed approximately identical results.

Results on concrete prisms, presented in Figure 9, showed that concrete with GF develops less expansion than control concrete. At one year of curing under test conditions, the control concrete with and without additionnal NaOH developed expansions of 0.22 % and 0.18 %, respectively, which according to CSA 23.2-27A, confirms the high reactivity of the Spratt aggregate used (> 0.12 %). In the concrete with glass frit, despite its high alkali-content, expansions of 0.092 % and 0.061 % were obtained. Since the same Spratt aggregate was used, it obviously means that GF has reduced the expansions.

The remakable performance of GF on the mechanical properties of the mortar implies a good pozzolanic reactivity of GF. It is generally admitted that most pozzolanic materials have an effectiveness in reducing the alkali-silica reaction (CSA A23.2-27A). Mineral admixtures contribute in transforming the portlandite $Ca(OH)_2$ released by portland cement during hydration into a C-S-H. Thus, the OH^- ions concentration in the interstitial solution which controls expansion is reduced (10, 11, 12). Also, among many other hypotheses concerning the mechanisms by which the use of mineral admixtures contributes to the control of alkali-silica reaction, is the probable insoluble alkalies present in most mineral admixures which therefore reduces the hydroxyl ion concentration of the pore solution. Chemical analysis of the solution extracted from the paste containing 0, 25 and 45 % cement replacement at 5, 60 and 120 minutes of hydration (Figure 10) showed this tendency. All the alkalis found in the solution are those released by the cement only. This may explain why expansion is lower for mortar bars and concrete prisms with GF. However, with regard to results on both mortar and concrete, the effect of curing temperature on expansions due the alkali-silica reaction is well established.

Further tests on mortars and concretes with different blending combinations and for longer periods of curing are currently on-going.

CONCLUSION

Adding glass frit to mortar or concrete substantially improves workability and strength. Workability increases as the rate of replacement increases. For a given slump, mortar or concrete containing glass frit requires less water or water reducer, which therefore can considerably enhance strength. At 28 days, the compressive strengths obtained on mortar containing GF are slightly higher than those of mortar incorporating the same percentage of slag and the control concrete. Similar results were also observed in concrete. Chloride ion permeability was significantly improved by incorporating glass frit as cement replacement. Despite its high alkalis content, glass frit mortar did not exhibit higher expansion. Further tests related to the durability of concrete are on-going.

REFERENCES

1. Wu, Zichao and Naik, Tarun R., "Properties of concrete produced from multicomponent blended cements", Cement and Concrete Research, Volume 32, N°12, December 2002, pp 1937-1942.

2. Hassan, K.E., Cabrera, J.G., and Maliehe, R.S., "The effect of mineral admixtures on the properties of high-performance concrete", Cement and Concrete Composite, Volume 22, 2000, pp. 267-271.

3. Wee, Tiong Huan and Matsunaga, Y., "Production and properties of high strength concretes containing various mineral admixtures", Cement and Concrete Research, Volume 25, N° 4, may 1995.

4. Malhotra, V.M, Mehta, P.K., "Pozzolanic and Cimentitious Materials", Advances in Concrete Technology, Volume 1, Gordon and Breach Publishers, 1996.

5. Groupe International de Contact Statistique sur l'Aluminium - Mandat, Industrie Canada, diffusé le 13.06.2002.

6. Kantro, D.L., « Influence of water Reducing Admixtures on properties of Cement Paste- A Miniature Slump Test", Cement, concrete and aggregate, CCAGDP, V.2, 1980, pp.95-102.

7. Smolczyk, H.G. « Slag structure and identification of slag » ; Proceedings, Seventh International Congress on the Chemistry of Cement 1 Paris; 1980.

8. Takayuki Kojima, Nobuaki Takagi and Kensaku Haruta; "Expanding characteristics of mortar with glass powder produced from waste bottles", XI International Conference on Alkali-Aggregate Reaction, Quebec 2000.

9. Hudec, Peter P. and Ghamari, R. Cyrus, "Ground waste glass as an alkali-silica reactivity inhibitor", XI International Conference on Alkali-Aggregate Reaction, Quebec 2000.

10. Shehata, M.H. and Thomas, M.D.A., "Use of ternary blends silica fume and fly ash to supress expansion due to alkali-silica reaction in concrete." Cement and Concrete Research, Volume 32, N° 3, March 2002 pp 341-349.

11. Shehata, M.H. and Thomas, M.D.A., "The effect of fly ash composition on the expansion of concrete due to alkali-silica reaction", Cement and Concrete Research, Volume 30, N° 7, July 2000, pp. 1063-1072.

12. Ramlochan, T., Thomas, M., and Gruber, K.A., "The effect of metakolin on alkali-silica reaction in concrete," Cement and Concrete Research, Volume 30, N° 3, March 2000, pp. 339-344.

Recycling Concrete and Other Materials

Table 1. Mix design of concrete

Material	Control concrete	Concrete with glass frit	Concrete with blast furnace slag
Cement (kg/m^3)	350	262.5	262.5
GF (kg/m^3)	0	87.5	0
Slag (kg/m^3)	0	0	87.5
Water (kg/m^3)	157.5	157.5	157.5
Sand (kg/m^3)	756	748	750
Aggregate 14 mm (kg/m^3)	484	484	484
Aggregate 20 mm (kg/m^3)	585	585	585
Air-entraining agent (ml/100kg binder)	50	60	60
Water reducer (ml/100kg binder)	228.6	114.3	228.6

Table 2. Chemical composition of GF

Components	Percentage by mass (%)		
	Type 10 cement	Glass frit	Blast furnace slag
SiO_2	19.9	31.7	36.8
Al_2O_3	4.2	23.4	10.32
Fe_2O3	3.4	3.4	0.73
CaO	62.3	14.6	36.54
CaF_2	---	9.4	---
MgO	2.0	0.76	12.56
Na_2O_{equi}	0.86	10.0	0.66
SO_3	0.00		0.24
L.O.I	<1.0		0

Table 3: Compressive strength in mortar of constant slump

Mortar	W/B	Compressive strength			
		1 d	7d	28d	100d
Control	0.485	16	35	47	52
20% GF	0.46	13.5	32	46.5	56
20% GGBFS	0.47	11	29	43	49

Table 4: Activity index

Activity index	Activity index (%)		
	7 d	28 d	100 d
CSA specification	75	75	Not given
Mortar with 20% GF	92	99	107
Mortar with 20% GGBFS	84	92	94

Table 5. Ratio of glass frit strength to slag strength

	Strength of glass frit mortar / Strength of slag mortar		
	10% replacement	25% replacement	45% replacement
1	0.89	0.97	0.90
7	1.03	1.0	1.24
28	0.86	0.97	1.08
91	0.88	0.92	0.99

Table 6. Characteristics of fresh concrete

Parameters	Hydration time	Control concrete	Glass frit concrete	Slag concrete
Slump (mm)	10 min	100	110	105
	30 min	70	80	70
	60 min	60	60	50
Air content (%)	10 min	5.8	5.5	5.3
	30 min	4.8	4.7	4.8
	60 min	4	4.4	4.1
Bulk density (kg/m^3)	10 min	2330	2290	2330
	30 min	2360	2350	2360
	60 min	2380	2360	2360
Temperature (°C)	10 min	21.9	21.7	20.8
	30 min	22.4	21.6	20.6
	60 min	22.3	21.5	21.1

Table 7. Initial and final setting times of concretes

	Initial setting	Final setting
Control	6 h 20 min	7 h 30 min
GF	6 h 30 min	7 h 50 min
GGBFS	6 h 45 min	8 h 15 min

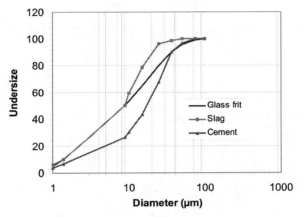

Fig. 1. Particle size distribution

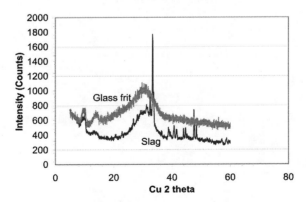

Fig. 2 Spectrum of GF and GGBFS

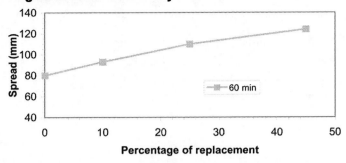

Fig. 3. Variation of the slump of the grout with the percentage of replacement

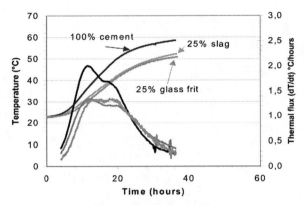

Fig. 4. Heat of hydration of the paste

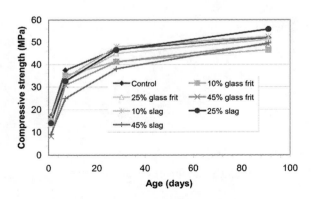

Fig. 5. Compressive strength variation (water to binder ratio = 0.45 constant)

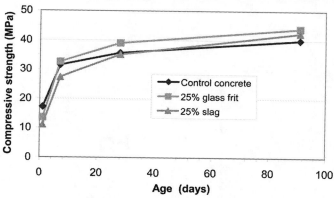

Fig. 6 Compressive strength variation with age

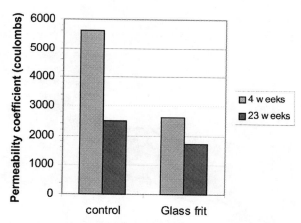

Fig. 7 Chloride ion permeability in concrete

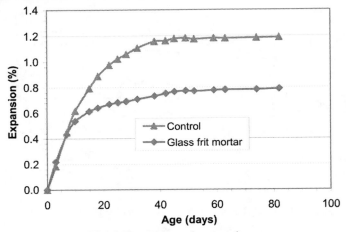

Fig. 8 Expansion of mortar bars

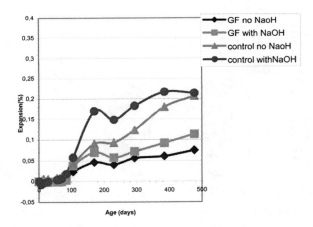

Fig. 9 Expansion on concrete prisms

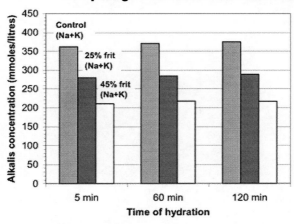

Fig. 10 Alkalis concentration in extracted solution

CONVERSION FACTORS—INCH-POUND TO SI (METRIC)*

To convert from	to	multiply by
Length		
inch	millimeter (mm)	25.4E†
foot	meter (m)	0.3048E
yard	meter (m)	0.9144E
mile (statute)	kilometer (km)	1.609
Area		
square inch	square centimeter (cm^2)	6.451
square foot	square meter (m^2)	0.0929
square yard	square meter (m^2)	0.8361
Volume (capacity)		
ounce	cubic centimeter (cm^3)	29.57
gallon	cubic meter (m^3)‡	0.003785
cubic inch	cubic centimeter (cm^3)	16.4
cubic foot	cubic meter (m^3)	0.02832
cubic yard	cubic meter (m^3)‡	0.7646
Force		
kilogram-force	newton (N)	9.807
kip-force	newton (N)	4448
pound-force	newton (N)	4.448
Pressure or stress (force per area)		
kilogram-force/square meter	pascal (Pa)	9.807
kip-force/square inch (ksi)	megapascal (MPa)	6.895
newton/square meter (N/m^2)	pascal (Pa)	1.000E
pound-force/square foot	pascal (Pa)	47.88
pound-force/square inch (psi)	kilopascal (kPa)	6.895
Bending moment or torque		
inch-pound-force	newton-meter (Nm)	0.1130
foot-pound-force	newton-meter (Nm)	1.356
meter-kilogram-force	newton-meter (Nm)	9.807

162 Conversion Factors—Inch-Pound to SI (Metric)

To convert from	to	multiply by
	Mass	
ounce-mass (avoirdupois)	gram (g)	28.34
pound-mass (avoirdupois)	kilogram (kg)	0.4536
ton (metric)	megagram (Mg)	1.000E
ton (short, 2000 lbm)	megagram (Mg)	0.9072
	Mass per volume	
pound-mass/cubic foot	kilogram/cubic meter (kg/m^3)	16.02
pound-mass/cubic yard	kilogram/cubic meter (kg/m^3)	0.5933
pound-mass/gallon	kilogram/cubic meter (kg/m^3)	119.8
	Temperature§	
deg Fahrenheit (F)	deg Celsius (C)	$t_C = (t_F - 32)/1.8$
deg Celsius (C)	deg Fahrenheit (F)	$t_F = 1.8 t_C + 32$

* This selected list gives practical conversion factors of units found in concrete technology. The reference source for information on SI units and more exact conversion factors is "Standard for Metric Practice" ASTM E 380. Symbols of metric units are given in parentheses.

† E indicates that the factor given is exact.

‡ One liter (cubic decimeter) equals 0.001 m^3 or 1000 cm^3.

§ These equations convert one temperature reading to another and include the necessary scale corrections. To convert a difference in temperature from Fahrenheit to Celsius degrees, divide by 1.8 only, i.e., a change from 70 to 88 F represents a change of 18 F or 18/1.8 = 10 C.

Index

A
aggregate, 77
alkali silica reaction (ASR), 61, 77
architectural concrete, 77
asphalt, 35
autogenous shrinkage, 115

B
bases, 35
bleed water, 85
bricks, 35
buried infrastructure, 99

C
cement, 85
chloride ions, 131
Chun, Y., 85
coal fly ash, 115
compressive strength, 85
concrete, 1, 35, 131, 145
concrete recycling, 47, 61
concrete waste, 11
construction and demolition waste, 1, 35
construction waste, 11
controlled low-strength materials (CLSM), 85

D
demolition, 1, 35
demolition debris, 47
drying shrinkage, 115
durability, 61

F
factorial experiments, 99
flowability, 115
flowable slurry, 85

fly ash, 61

G
glass, 77
glass frit, 145
Gress, D. L., 61
ground granulated blast furnace slag, 61
guidance, 47

H
Hansen, T. C., 35

I
integrated waste management, 1

J
Japan, 11

K
Kasai, Y., 11
Khan, A., 99
Kraus, R. N., 85

L
Laldji, S., 145
latex paint, 131
Lauritzen, E. K., 1, 35
limestone dust, 115
lithium nitrate, 61

M
Melton, J. S., 47
Meyer, C., 77
mineral admixture, 145

N
Naik, T. R., 85

Nehdi, M., 99, 131

P
permeability, 85

R
recycled aggregate, 11
recycled concrete, 11
recycled concrete aggregate, 47, 61
recycling, 1, 35, 47, 77, 99, 131
Riefler, C., 115
road construction, 35

S
Scott IV, H. C., 61
segregation, 115
self-consolidating concrete, 115
Shi, C., 115
Shimanovich, S., 77
Siddique, R., 85
sidewalks, 131
silica fume, 61
specification, 47
spent pot liner, 145

strength, 115
stress-strain curve, 99
sub-bases, 35
surface scaling, 131
sustainable development, 77, 145

T
Tagnit-Hamou, A., 145
tire rubber, 99
tri-axial compression, 99
tunnel linings, 99

U
uni-axial compression, 99

W
waste, 131
wood ash, 85
Wu, Y., 115